Palgrave Studies in Adaptation and Visual Culture

This series addresses how adaptation functions as a principal mode of text production in visual culture. What makes the series distinctive is its focus on visual culture as both targets and sources for adaptations, and a vision to include media forms beyond film and television such as videogames, mobile applications, interactive fiction and film, print and nonprint media, and the avant-garde. As such, the series will contribute to an expansive understanding of adaptation as a central, but only one, form of a larger phenomenon within visual culture. Adaptations are texts that are not singular but complexly multiple, connecting them to other pervasive plural forms: sequels, series, genres, trilogies, authorial oeuvres, appropriations, remakes, reboots, cycles and franchises. This series especially welcomes studies that, in some form, treat the connection between adaptation and these other forms of multiplicity. We also welcome proposals that focus on aspects of theory that are relevant to the importance of adaptation as connected to various forms of visual culture.

Pamela Demory
Editor

Ecoadaptation

Mediating Nature and the Environment

palgrave
macmillan

Editor
Pamela Demory
Eugene, OR, USA

ISSN 2634-629X ISSN 2634-6303 (electronic)
Palgrave Studies in Adaptation and Visual Culture
ISBN 978-3-032-09846-7 ISBN 978-3-032-09847-4 (eBook)
https://doi.org/10.1007/978-3-032-09847-4

This Palgrave Macmillan imprint is published by the registered company Springer Nature Switzerland AG
The registered company address is: Gewerbestrasse 11, 6330 Cham, Switzerland

Acknowledgements

The central theme of this collection grew out of the 2023 Literature and Film Association conference in Montana: Ecologies of/and Adaptation, and so my first thanks must go to the board members of the Literature/Film Association for putting together such a wonderfully thought-provoking conference, Marty Marko for providing an intellectually stimulating and congenial environment, Allen Redmon for helping me to sort out the initial concept, and all the presenters, whose ideas sparked my imagination. Despite those sparks, this book would never have come to be had Julie Grossman not turned to me at one point during the conference and said, "You should really put together an edited collection on this topic." Since that moment two years ago, Julie has been an unflagging and generous supporter of this project, and I cannot thank her enough for her editorial advice and cheerful communications. I must also thank all the contributors whose wonderful ideas, patience with my nitpicky editing, and enthusiasm for the project have made the work on this book not only intellectually engaging but also great fun. At Palgrave, my thanks go to Publishing Editor Lina Aboujieb and Production Editor Shruthi Chelliah Mariappan, who guided me with gentle professionalism through the entire process. I also have many people to thank for general writing inspiration and support: my writing group, the Nonfictional Friends—Helen, Brenda, Brianna, John, Beth, and Michelle; Sondra Reid, my colleague, friend, and editor extraordinaire; and my mother, Maxine, who taught me by example that a regular writing regimen is a key to a happy retirement. And finally, I am grateful to extended family members who generously gave me space and time to work and plenty of emotional support: Mark, Carrie, Deborah, Paula, my mother (again), and, especially, Kent and Coal and Vera.

Contents

Notes on Contributors

Petr Bubeníček is Associate Professor at the Faculty of Arts of the Masaryk University in Brno, Czech Republic. He specializes in adaptation studies, intermedial studies, ecocriticism, and modern Czech literature and culture. Additionally, he is the guarantor of the Literature and Intercultural Communication study program, which emphasizes literature, media, and culture. He published a book, *Subversive Adaptations: Czech Literature on Screen Behind the Iron Curtain* (2017). Together with Tereza Dědinová, he coedited an ecocritical issue of the scholarly journal *Česká Literatura* (2023). He is currently writing the book *The Trap of the Everyday: Czech Literature on Screen 1970–1995*.

Christopher R. Carter is Profesor Ayudante Doctor at the Universidad de Castilla–La Mancha in Ciudad Real, Spain, where he teaches in the English Studies program. He obtained his Ph.D. in Spanish Literature from the University of Nebraska–Lincoln, and his research focuses on the intersections of gender, sexuality, translation, and adaptation. He has published translations in the *North American Review* and research in Routledge's *Cultural Representations of Queer Aging in Contemporary Spain* anthology.

Cat Champney is currently a Ph.D. candidate at the University of Delaware. Her dissertation project explores adaptation(s) of the Bluebeard fairy tale and the female Gothic.

Augustto Correa Cipriani is Lecturer of Brazilian Literature at Universidade Federal de Minas Gerais (UFMG). He holds a Doctorate and a Master's degree in Literary Studies from UFMG. His research interests include Portuguese-speaking literatures and literary theory, with a focus on word and image studies and the relationship between poetry and painting. He is a researcher at "*Intermídia: Estudos sobre a Intermidialidade*" (CNPq/UFMG) and has presented conference papers on the history of graffiti in Brazil as well as on contemporary Indigenous art and literature. Additionally, he has published articles on poetic graffiti in Brazil and visual poetry.

Pamela Demory is Lecturer Emerita from the University of California at Davis. She has worked in the field of adaptation studies for thirty years. Her articles and book chapters include analyses of queer adaptation, *Dickinson, Moonlight, Hedwig and the Angry Inch, Queer as Folk, Twilight, Pride and Prejudice, Wise Blood, and Apocalypse Now*. She is on the editorial boards of *The Journal of Popular Culture* and *Adaptation.* She has edited two volumes of critical essays: *Queer Love in Film and Television* (with Christopher Pullen, 2013) and *Queer/Adaptation* (2019). Her current writing projects include explorations of nature writing and adaptation.

Tereza Dědinová is Assistant Professor at the Faculty of Arts of the Masaryk University in Brno, Czech Republic. She focuses on speculative fiction from cognitive and ecocritical perspectives. Her recent projects include coedited volumes *Images of the Anthropocene in Speculative Fiction*, with Weronika Łaszkiewicz and Sylwia Borowska-Szerszun (2021), and *Fantasy and Myth in the Anthropocene*, with Marek Oziewicz and Brian Attebery (2022). Together with Petr Bubeníček, she coedited an ecocritical issue of the scholarly journal *Česká Literatura* (2023). Currently, she is participating in a project focused on effective environmental communication within the nonprofit organization Institut 2050.

Robert Geal is Senior Lecturer in Film and Media Studies at the University of Wolverhampton, UK. His research interests include ecocriticism, psychoanalysis, and adaptation studies, and in particular the intersection of these subjects and methods. He is author of the monographs *Anamorphic Authorship in Canonical Film Adaptation* and *Ecological Film Theory and Psychoanalysis: Surviving the Environmental Apocalypse in Cinema*, as well as numerous articles and chapters including "Towards an Ecocritical Adaptation Studies," which won the journal *Adaptation*'s 2022 Early Career Researcher Prize.

Thomas Grochowski is Professor and Chair of English at St. Joseph's University, New York, where he regularly teaches courses on film, literature, and adaptation. He serves on the editorial board of *Literature/Film Quarterly*, the *Journal of Film and Video*, *Popular Music and Society*, and *Rock Music Studies*. He is also a longtime member of the board of the Mid-Atlantic Popular and American Culture Association (MAPACA). His many article publications include adaptation-oriented essays on *I'm Not There* and *The Last Waltz*. This chapter was completed with the help of a Faculty Small Grant from St. Joseph's University, New York.

Elizabeth D. Gruber is Professor of Languages and Literatures at Commonwealth University in Pennsylvania. She has published on early-modern literature, Shakespearean adaptations, and ecocriticism. Her recent book, *The Eco-Self in Early Modern English Literature*, argues that the period hosted the birth of a new subset of the human, melding a deeply introspective turn with awareness of human's embedment in the world. Current research addresses the promising alliances among adaptation studies, ecocriticism, and cognitive theory. Focusing on *The Tempest* and several adaptations of Shakespeare's play, this project harmonizes biological/material concerns with psychological theories of the self, showcasing the benefits of an ecophenomenological approach.

Juliana Jones-Beaton is currently a Ph.D. candidate in literature at the University of Delaware, studying fungal networks, ecosystems, and the environment in post-2001 speculative fiction. Her dissertation is provisionally titled "A Body that Fruits: The Fungus in Contemporary Speculative Fiction." She has previously published in *Ampersand, Public Seminar, Moss Puppy Magazine,* and *Failed Architecture.*

Katie Kane was raised in North Dakota and now lives in Missoula, Montana. An Associate Professor in the English Department at the University of Montana, Kane has published her scholarship on colonialism, capitalism, and culture in journals such as *American Indian Quarterly*, *Cultural Studies*, and *Cultural Critique*. She is at work on a manuscript called *The Deep North: Indigeneity, Big Oil, and Resistance* that investigates the northern region of the North American Rocky Mountain West. Kane has published short stories and autofiction in *Black Warrior Review*, *Salvage, Fence Magazine*, and *Twelve Winters.*

Susan Kollin is Director of American Studies and Professor of English at Montana State University–Bozeman. Her research and teaching areas include ecocriticism and the environmental humanities, western American fiction and film, and feminist theory. She is the author of *Thelma & Louise*, published by the University of New Mexico Press and is a past president of the Western Literature Association.

Thomas Leitch recently retired from the University of Delaware, where for many years he directed the Film Studies program. His most recent books are the edited collection *The Scandal of Adaptation* and the textbook *Encounters with Adaptation.*

Clara Matheus Nogueira is Professor of English Literature at Universidade Federal de Campina Grande (UFCG). She holds a Doctorate and a Master's in Literary Studies from Universidade Federal de Minas Gerais (UFMG). A member of the research group "*Intermídia: Estudos sobre a Intermidialidade*" (CNPq/UFMG), Clara has taught and published on intermedial studies and literatures in English. Her research focuses on digital media, especially adaptations of William Shakespeare's plays to social media. She cofounded *mimimidias*, a digital academic outreach project in Brazil.

Matthew Page is an independent scholar who has been specializing in the cinematic adaptation of biblical narratives for nearly twenty-five years. His book *100 Bible Films* was published by the British Film Institute in 2022. He has also contributed to numerous edited collections. Recent chapters include "*Il Vangelo secondo Matteo*" (2023), "Jesus of Cinecittà" (2021), "The Nativity Re-Born: Genre and the Birth and Childhood of Jesus" (2020), and "'Can We Try that Again?' The Fate of the Biblical Canon on Film" (2018). He previously worked as an environmental consultant.

Christina Parker-Flynn is Associate Professor of Film and Literature at Florida State University. She holds a Ph.D. in Comparative Literature from Emory University and specializes in film theory and adaptation studies. She is the author of *Artificial Generation: Photogenic French Literature and the Prehistory of Cinematic Modernity* (2021), which investigates the intersection of film theory and nineteenth-century literature, arguing that the depth of amalgamation that occurred within literary representation during this era aims to replicate an illusion of life and its sensations, in ways directly related to broader transitions into our modern cinematic age.

Agnieszka Rasmus is an Adjunct Professor in the Department of Studies in English Drama, Theatre and Film at the University of Lodz, Poland, where she teaches literature and film. Her published work focuses on adaptation and seriality in popular culture, with specific interest in Hollywood, British cinema, and Shakespeare. Her book *Hollywood Remakes of Iconic British Films: Class, Gender and Stardom* (EUP, 2022) was awarded the prize for the best transnational studies book of 2022.

Allen H. Redmon is Professor of English and Film Studies at Texas A&M University–Central Texas. He is the author of *Rewatching on the Point of the Cinematic Index* (University Press of Mississippi, 2022) and *Constructing the Coens: from* Blood Simple *to* Inside Llewyn Davis (Rowman & Littlefield, 2015). He also is the editor of *Next Generation Adaptation: Spectatorship and Process* (University Press of Mississippi, 2021) and co-editor of *Clint Eastwood's Cinema of Trauma* (McFarland, 2017). He is a past president of the Literature/Film Association and a board member of the Association of Adaptation Studies.

John Sanders is Assistant Professor of English at Appalachian State University, where he teaches classes on film, television, and new media. His work is situated at the interstices of games, literature, and narrativity, and has appeared in journals and fora such as *gamevironments, Adaptation,* and PAX East. He also serves as a board member and web master for the Literature/Film Association, of which he has been an active member since 2017. He happily resides in the mountains of North Carolina with his partner and their two dogs: Henry and Mr. Darcy.

Kim Waale is Artist-in-Residence and Visiting Professor in the Visual and Performing Arts Department at Le Moyne College, New York. Waale has been awarded grants, fellowships, and artist residencies from organizations including I-Park (environmental art); Sculpture Space, Light Work (photography), International Studio and Curatorial Program, Saltonstall Foundation, Harlech Biennial (Wales, UK) and A Solaina de Piloño (Galicia, Spain). She exhibits in art venues such as Stone Quarry Art Park, Plas Glyn-y-Weddw (Wales, UK), Hallwalls Contemporary Art Center, A. I. R. Gallery, McCormick Freedom Museum, Housatonic Museum, Memorial Art Gallery, and the Everson Museum. Waale's scholarship focuses on adaptations within visual art.

List of Figures

1

What is Ecoadaptation? And Why Should We Care?

Pamela Demory

This collection of essays brings together the study of cultural adaptations and environmental studies (or ecocriticism)—thus the coined term "ecoadaptation"—to explore how elements of the natural world and the environment are represented and adapted in a variety of media. Note that I refer here to both "ecocriticism" and "environmental studies," not because they are interchangeable, but because this collection encompasses both, and the two terms do overlap. Ecocriticism tends to be used in connection with literature and cultural studies, whereas environmental studies tends to include the social sciences and natural sciences as well. But the lines between the two are blurry, and both can be described as interdisciplinary explorations of the relationship between human culture and the natural environment.

This is the first book-length work to explore the intersection of adaptation studies and environmental studies. It encompasses a remarkable

P. Demory (✉)
Eugene, OR, USA
e-mail: phdemory@ucdavis.edu

P. Demory (ed.), *Ecoadaptation*, Palgrave Studies in Adaptation and Visual Culture,
https://doi.org/10.1007/978-3-032-09847-4_1

diversity of theoretical approaches—including psychoanalysis, intermedial studies, Indigenous studies, ecofeminism, ecophenomenology, and genre theory. It studies a range of media—from ancient texts to classic and contemporary novels; independent, genre, and documentary film; radio, television, music, art, graphic novels, graffiti, and video games. And it has geographic breadth—analyzing texts authored in or set in the United Kingdom, the United States, the Czech Republic, Mexico, Spain, Brazil, Poland, Africa, and the Caribbean. It includes work from established scholars in adaptation studies, as well as scholars in the fields of literature, film, art, and media studies.

The central theme of the collection grew out of the 2023 Literature and Film Association conference in Montana: Ecologies of/and Adaptation, and several of the articles included here are expanded versions of papers presented at this conference. It is also inspired by the work of recent scholars in adaptation studies who have identified the current moment of environmental crisis and climate change as an important opportunity for new explorations of the ways that adaptation studies and ecocritical studies might inform each other. Kyle Meikle, for example, calls for a "green movement in adaptation studies" (2020). Robert Geal argues persuasively that adaptation studies can help us to understand "fundamental ecocritical question[s]... by outlining how human attitudes towards the environment, and cultural expressions of those attitudes in the form of various kinds of texts, are adapted through time, across cultures, between texts" (2023). And Tomas Elliott argues that "adaptation studies... seem ideally situated to interrogate [representational] strategies, assessing how certain cultural forms, produced under specific social, political, and material conditions, process and represent the similarities and differences between adaptation in biology and adaptation in art" (2024). Together, these scholars have mounted a challenge for adaptation studies scholars that this collection responds to.

Why adaptation? What do adaptation scholars (specifically) offer to the discussion of environmentalism (that other media scholars don't)? This is the foundational question upon which the entire collection rests. To answer this question, I begin with history. The fields of adaptation studies and ecocriticism have developed in parallel ways. Ecocriticism began as a field focused specifically on the relationship between literature and the environment and eventually broadened to become an interdisciplinary study of the many interrelationships between human and nonhuman (or more-than-human) natures, environmentalism, and climate change. Adaptation studies, in a similar trajectory, developed from a narrow focus on the relationship between literature and film to become an interdisciplinary field attentive to the many and varied ways that adaptation functions in all kinds of cultural texts.

Both fields took a poststructuralist turn at the beginning of the twenty-first century (represented perhaps most notably by Dana Phillips and Timothy Morton in ecocriticism and Robert Stam in adaptation studies), challenging the binary oppositions of nature/culture and adaptation/original in their respective fields, and emphasizing the role of language, of text, of mediation in both fields. In both fields, scholars noted that the attempt to faithfully represent the original source (whether that was a landscape or a tree, in one case, or a Shakespeare play or a novel, on the other) was both impossible and undesirable. "Faithful rendering" of nature and "fidelity" to an original became hallmarks of an outmoded scholarship.

Later in the twenty-first century, both fields again turned, this time away from a linguistic focus to materialism. Ecocritical scholars such as Stacy Alaimo focused on questions of the body; Simone Murray established a new focus in adaptation studies on the adaptation industry. In both fields, scholars paid new attention to the agency of the nonhuman world and to the agency of stories. The fields became newly entangled. Ecological language became more common in adaptation studies, with Simone Murray describing the adaptation industry as an "ecosystem" (24), and Thomas Leitch referring to "an ecology of textual studies" ("Against Conclusions" 700). And in ecocriticism, the role of narrative became more prominent, with scholars Iovino and Opperman, for example, asserting that "the world's phenomena are knots in a vast network of agencies which can be 'read' and interpreted as forming narratives, stories.... all matter, in other words, is a 'storied matter'" (1). The language of interdependence, mesh, network, and web are now common to both fields.

Bortolotti and Hutcheon established the now-common analogy between biological and cultural evolution in their influential 2007 study. But connections between the two fields have existed from early on. The environment is already inherently a place of adaptation. Frederick Turner, in 1991, describes evolution as "the *interference* of everything with everything else" (43), which also might serve as a definition of adaptation studies. Cheryll Glotfelty, in her influential introduction to ecocriticism in the 1990s, defines ecocriticism as the "*relationship* between literature and the physical environment," an idea echoed more recently by Dennis Cutchins in his introduction to the *Routledge Companion to Adaptation Studies,* where he defines adaptation as being fundamentally about the *relationship* between texts. And Lawrence Buell, in another influential ecocritical text from the 1990s, *The Environmental Imagination*, notes that the natural environment as we understand it has been constructed in successive cultural movements into an "ideological palimpsest," an image that has become common in adaptation studies since Linda Hutcheon described adaptation, in a much-quoted line, as "secondary without being secondary. It is its own palimpsestic thing" (9).

To return to the initial question, "Why adaptation?" I turn to Timothy Corrigan's definition of adaptation in the *Oxford Handbook of Adaptation Studies*, as the study of "evolution, fidelity, and intertextuality" (24). It is the intertwining of these three concepts that makes adaptation studies distinct from other media studies and which provides adaptation studies scholars the tools for studying environmental texts. Further, the concept of the adaptation system as an "ecosystem" moves the biological Darwinian metaphor about adaptation to a system-wide environmental concept that intertwines the two fields even more tightly.

So what is "ecoadaptation"? The term refers, first of all, to a text that is an adaptation and is also "environmental" in some way. It might be an adaptation that refocuses our attention on the environment, or that foregrounds the relationship between the human and nonhuman (or more-than-human) world. Or one that enables us to see environmental issues or that records changes over time in our attitudes to or our relationship with nature. But "adaptation" is both a product and a process, and so ecoadaptation describes not only a particular kind of text but also a way of conceptualizing a complex process of meaning-making and transformation. Ecoadaptation might refer to our ability to adapt ourselves—via narratives, art, cultural work—to climate change. It might refer to the sustainability of the technologies we use to reproduce our stories. It might also describe how a text, or organism, adapts to a new ecosphere. It may describe not so much a *thing* as a critical lens, a way of reading. Ecoadaptation studies, to extend the concept even further, could refer to the study of how texts or organisms or other materials adapt to a new or changed ecosphere, one that includes human contexts but also a broader range of ecological contexts. Ecoadaptation, then, implies the consideration of a complex ecosystem of texts and contexts.

This book is organized into five sections that explore key ecoadaptation questions:

I. **The Nature of (Eco)Adaptation**. What is the nature of adaptation? How is cultural adaptation related to biological adaptation? What theoretical frameworks might help to explicate the concept of ecoadaptation?
II. **Adaptation in the Age of Climate Change**. How do environmentally focused adaptations encourage/discourage or mediate public response to environmental crisis—including climate change? What specific forms/media best represent environmental concerns in the Age of the Anthropocene?
III. **Indigenous Nature and Culture**. How have Indigenous cultural narratives about nature and the environment been transformed or appropriated? What has the legacy of colonialism and white settler culture meant for the health of the planet?
IV. **The Nature of Genre Adaptation**. How do genres represent and adapt environmental themes? How might genres adapt to explore the very concept of nature?
V. **Adapting Nature**. How have elements of the natural world itself been adapted, represented, replicated in various art forms—narrative and non-narrative?

Although each of these sections provides a coherent focus, most chapters address more than one of these questions, and readers will find many filaments or threads that connect chapters across sections. For example, ecofeminist readings connect "The Gloom in the Attic" in Section I with "*Indigo*'s Challenge to *The Tempest*" in Section III; biblical texts connect

"Is Adaptation Natural?" in Section I with "The Ark of the Moral Universe" in Section II; the American West connects "A Huge Mass in a Single Hand" in Section III with "Settler Ecologies and Western Adaptation" in Section IV; environmental art connects "This is Indigenous Land" in Section III with "Material Worlds" in Section V; Shakespeare connects "Posters of the Sea and Land" in Section II with "*Indigo*'s Challenge" in Section III; representation of environmental crises connect "A History of Urgency" in Section I with "Geo-Ecolocating and Bio-Boomeranging" in Section II; avian research connects "Settler Ecologies" in Section IV with "Biocinemimicry in *The Birds*" in Section V; questions about *human* nature connect "I Can Hear the Glaciers Melting" in Section II with "Against Nature?" in Section IV. These are only a few of many interconnections that attentive readers are likely to discover here.

All these chapters explore in one way or another these central questions: How is our conception of nature—or the environment more generally—framed by the stories we tell about it? And how do those stories evolve, mutate, and replicate over time and space and in different media?

In recent years, adaptation studies scholarship has tended to move away from case studies to consider larger structural, philosophical, and theoretical questions. But in his introduction to the *Oxford Handbook of Adaptation Studies*, Thomas Leitch argues that despite the excellent arguments for moving adaptation scholarship beyond the tired trope of book-to-film analysis, case studies are still useful for illustrating the implications of significant concepts in adaptation. Many of the essays in the collection are not, primarily, case studies. Petr Bubeníček and Tereza Dedinová describe their chapter as a "case study" of documentary Czech film, but it is actually a much more theoretically informed study that uses the Czech documentaries to make an argument about how adaptation, intermedial studies, and ecocriticism intersect. Other essays take elements of the environment as their primary object of analysis: Kim Waale on art installations that modify natural elements and work with natural adaptive processes; Christina Parker-Flynn on the interrelationship of bird physiology, avian-inspired mechanics, and Hitchcock's *The Birds*, among other cultural, environmental, and technical objects. Those contributors who do focus on case studies use them to explore significant implications for studying adaptation and environmentalism together. Champney and Jones-Beaton, for example, while focusing primarily on the relationship between Charlotte Perkins Gilman's *The Yellow Wallpaper* and *Mexican Gothic*, use that example to develop a theory of adaptation based on fungal networks. Chris Carter, in studying the relationship between a nineteenth-century Spanish novel and its twenty-first-century graphic novel adaptation, makes a case for the graphic novel itself as an ecologically important form. And Allen Redmon's analysis of the African film *Makibefo*, which adapts Shakespeare's *Macbeth*, is actually more interested in the material reality of the film production that witnesses climate change in action.

All together, this is a collection that takes adaptation studies much further away from the traditional text-to-text analysis than is common to adaptation studies collections. This is, still, a book firmly lodged in the realm of adaptation studies/cultural studies, and its perspective is culturally focused. I see this collection as a beginning, as a way of entering into what will no doubt become a rich field for scholars in both these broad fields to engage in further study, study that will no doubt include the perspectives of practicing ecocritics and environmental scientists.

The question of what good such a study will do is one that comes up in many ecocritical collections. Everyone agrees that scholarship should be tied to activism, that our collective aim is to actually make a difference in the world. Everyone also seems to agree that an academic book is probably unlikely to have a direct effect. Jørgen Bruhn and Niklas Salmose, for example, editors of the 2023 *Intermedial Ecocriticism: The Climate Crisis Through Art and Media*, ask "what difference to our crisis-ridden world will one more volume discussing different kinds of representations of the ecological crisis make?" Their answer, which I would echo, is that the purpose of an academic book is to provide methods for understanding, to help readers "to navigate better in, and be better informed about, the ongoing info-wars that are being waged concerning the planetary state of affairs" (11). In this collection, several chapters focus on the way adaptations *reflect* public attitudes but probably don't *effect change* (e.g., Geal, and Page, on adaptations of the Noah story; Rasmus on subverting the genre of the zombie film). Others show how certain adaptation strategies (e.g., Bubeníček and Dedinová on environmental documentaries, Redmon on the Madagascar adaptation of *Macbeth*, and Waale on environmental art) reveal climate change in action. Other chapters (e.g., Carter; Grochowski; Sanders) offer the possibility that different forms of cultural expression—the graphic novel, radio, videogame—might help people understand climate change (or environmental degradation) in a way that traditional media do not. Cipriani and Nogueira provide one example of activist adaptation that is meant to challenge public perceptions and to provoke policy change. And John Sanders provides an example of a videogame adaptation that, surprisingly, urges its players to go outside and walk in nature.

A book like this is where ecocritical education starts—with scholars who engage with the complex questions raised here and then use them in their classes where students learn new ways of thinking about the environment and the role of adaptation in both culture and science. It's a slow process but a significant one.

References

Alaimo, Stacy. *Bodily Natures: Science, Environment, and the Material Self*, Indiana UP, 2010.

Bortolotti, Gary R., and Linda Hutcheon. "On the Origin of Adaptations: Rethinking Fidelity Discourse and 'Success'—Biologically." *New Literary History*, vol. 38, 2007, pp. 443–458.

Bruhn, Jørgen, and Niklas Salmose. Introduction. *Intermedial Ecocriticism: The Climate Crisis Through Art and Media,* Lexington Books, 2024, pp. 1–13.

Buell, Lawrence, *The Environmental Imagination: Thoreau, Nature Writing, and the Formation of American Culture,* Belknap Press of Harvard UP, 1995.

Corrigan, Timothy. "Defining Adaptation." *The Oxford Handbook of Adaptation Studies,* edited by Thomas Leitch, Oxford, 2017, pp. 23–35.

Glotfelty, Cheryll. "Introduction." *The Ecocriticism Reader: Landmarks in Literary Ecology*. U of Georgia P, 1996. xv–xxxvii.

Iovino, Serenella, and Oppermann, Serpil. "Introduction: Stories Come to Matter." *Material Ecocriticism: Materiality, Agency, and Models of Narrativity,* Indiana UP, 2014, pp. 1–17.

Leitch, Thomas. "Against Conclusions: Petit Theories and Adaptation Studies." *The Oxford Handbook of Adaptation Studies*, edited by Thomas Leitch, Oxford UP, 2017, pp. 698–709.

Leitch, Thomas. "Introduction." *The Oxford Handbook of Adaptation Studies*, Oxford, 2017.

Morton, Timothy, *Ecology without Nature: Rethinking Environmental Aesthetics*. Harvard UP, 2007.

Murray, Simone. *The Adaptation Industry*. Routledge, 2012.

Phillips, Dana, *The Truth of Ecology: Nature, Culture, and Literature in America*. Oxford UP, 2003.

Stam, Robert. "Introduction: The Theory and Practice of Adaptation." *Literature and Film: A Guide to the Theory and Practice of Film Adaptation*, edited by Robert Stam and Alessandro Raengo, Blackwell, 2005, pp. 1–52.

Turner, Frederick. "Cultivating the American Garden." *The Ecocriticism Reader: Landmarks in Literary Ecology*, edited by Cheryll Glotfelty and Harold Fromm, U of Georgia P, 1996, pp. 40–51.

Part I

The Nature of (Eco)Adaptation

What is the nature of adaptation? How is cultural adaptation related to biological adaptation? What theoretical frameworks might help to explicate the concept of ecoadaptation?

The three chapters in this section pose three very different answers to these theoretical questions: a Lacanian psychoanalytic model as an alternative to the familiar biological evolutionary analogy; a strategy for integrating adaptation studies with intermedial studies and materialist ecocriticism; and a theory of adaptation as fungal network. Each of these chapters invites further exploration: What would a Lacanian signification theory tell us about adaptations outside a biblical context? What would an intermedial/materialist/adaptation study reveal about texts other than environmental documentaries? How might a fungal network theory help elucidate the intricate webs of another set of texts? Together, these chapters also raise intriguing questions about what *other* theories or combinations of theories might illuminate adaptation/ecocritical questions.

2

Is Adaptation Natural? Ecocritical Repercussions of Conceptualizing Adaptation as Memetic Evolution and/or as (Re)Iterations in the Signifying Chain

Robert Geal

This chapter considers some limitations of the biological-cultural analogy in (ecocritical) adaptation studies, and proposes Lacanian psychoanalysis as an alternative to the evolutionary account of textual adaptation. The chapter demonstrates the differences between these methods by considering how contemporary reworkings of the biblical "Noah" (*King James Bible*, Genesis 6.5–9.17) narrative keep or alter various aspects of the hypotext. This example has been chosen both because it demonstrates the difference between the approaches and because it directly relates to ecological issues.

R. Geal (✉)
University of Wolverhampton, Wolverhampton, UK
e-mail: R.Geal2@wlv.ac.uk

P. Demory (ed.), *Ecoadaptation*, Palgrave Studies in Adaptation and Visual Culture,
https://doi.org/10.1007/978-3-032-09847-4_2

Adaptations Relating to Cultural Environments or Signifying Chains

The application of ecocritical questions to adaptation studies highlights an already contested methodological issue in the study of adaptation—the extent to which textual adaptation is like biological adaptation. This issue is already contested because there had been debate about using biological analogies before such debates were explicitly related to ecological questions. When studies of adaptation do address explicitly ecological questions, the importance of the textual/biological relationship becomes more pressing, because textual/biological relationships are no longer merely methodological issues applied to non-ecological questions. So if we have a flawed textual/biological analogy that is used to explore questions about authorial fidelity, as in Gary Bortolotti and Linda Hutcheon's attempt to use "biological concepts in a heuristic manner to help free narrative adaptation theory from the conceptual restrictions of a misleading evaluative fidelity discourse" (454), then the potential repercussions are mostly constrained to conclusions about authorial fidelity. But if a flawed textual/biological analogy is used to explore questions about how textual adaptation relates to human attitudes toward the nonhuman environment, then the potential repercussions might include flawed conclusions about those human attitudes.

Tomas Elliott has recently argued that adaptation studies should contextualize the biological/textual analogy in order that we might "reveal and assess [our] contingent and changeable theories about nature and culture—such as, for example, the theory of evolution by natural selection or the spread of memes in a cultural environment" (333). This chapter positions the textual/biological relationship within the context of those broader "theories about nature and culture." That is, adaptation studies has tended to take a specific component of biology and apply this to a specific component of culture. This is exemplified in Bortolotti and Hutcheon's "homology between biological and cultural adaptation" (444), which develops Richard Dawkins' "concept of 'meme' as the cultural equivalent of the biological gene" (446–7) to argue that "stories, in a manner parallel to genes, replicate; the adaptations of both evolve with changing environments" (444). If we are to understand the specific textual/biological relationship between genes and memes, then it is necessary to contextualize it by addressing the underlying determinants which inform this specific relationship. We cannot hope to understand the relationship between textual adaptation and genetic adaptation if we do not first address the broader underlying phenomenon of human culture in relation to biology. We therefore need to ask whether and how human culture is a divergence from the rest of nonhuman nature.

A qualification to this question is immediately required. It should not imply a speciesist binary opposition, with humans and nature seen as distinct ontological categories—human culture arises out of natural conditions, with the species *homo sapiens* subject to the same physical laws as other aspects of nature. So the question should not imply that humanity is somehow *super*-natural, with humanity distinct from other aspects of nature because of unique nonmaterial factors like an immortal soul. Such biases, as Lynn White Jr. has shown, are culturally specific human inventions which facilitate anthropocentric exceptionalism. Nevertheless, eliding the issue of how human culture differs from the rest of nature may be just as perilous, with the potential to justify specifically cultural behavior such as racism, sexism, homophobia, inequality, war, and so on as entirely "natural."

There isn't room here to fully account for how existing ecocriticism addresses this issue, but there is certainly no definitive consensus that can be transposed into adaptation studies. In fact, the transhistorical and transcultural interests of adaptation studies may be one area where the field can assist this ecocritical lacuna. The model which I am proposing to address this lacuna is derived from Lacanian psychoanalysis, which explains how humanity's biological origins produce something unprecedented in the rest of nature.

Contrasting Cultural Environments with Signifying Chains

It makes sense to begin with the biological because, excepting a supernatural explanation, this is humanity's material substratum. I am taking it as axiomatic, then, that humans evolved from prior species according to the laws of evolution by natural selection. Part of our evolutionary specificity is a relatively large brain. It is not so much, however, that this size in and of itself produces an animal with capacities to create something like culture. Rather, the largeness of the human brain means that there is a relative disparity in size between the head and pelvis, so that human babies' heads are almost too large for them to be born. The evolutionary solution to this problem is for the physical development of the brain to continue after birth. As biologist Stephen Jay Gould claims, "at birth, our brains are still growing at fetal rates...; if this increase continued in utero, heads would soon become too big for successful parturition" (370). Thus, in contrast to the offspring of other mammals, which are very quickly able to directly engage with their environments, human babies are essentially born premature. The physical development of the human brain begins in the closed, solitary environment of the womb, but also continues in the open, communal environment of the social.

Catherine Malabou clarifies this biological specificity by utilizing the idea of epigenetics, which she describes as "the branch of biology that studies gene expression," i.e., "nongenetic changes or modifications" (104). Given human prematurity, our "brain's development... depends for a great part upon epigenetic factors. The anatomy of the brain is genetically determined. But innumerable synaptic connecting possibilities are not. Synapsis formation escapes genetic determinism" (104–5). To be clear, humans are no different from other animals in terms of being influenced by both genetic and epigenetic factors, and in this sense, humans are just as natural as other animals. What is different in humans, then, is not the unprecedented presence of epigenetic factors, but the relative importance of those epigenetic factors, which have an outsized significance compared to other animal species. So it is not the human per se which is exceptional. Rather, the distinctive exception relating to humans is a specific epigenetic response to both genetic and environmental conditions. Thus, for Adrian Johnston, "the genetically dictated prematuration of the human brain, in which much of maturation is left up to epigenetic dictates that follow birth, means that human nature is naturally destined for denaturalization" (77) or, as François Ansermet and Pierre Magistretti put it, humans are "genetically determined not to be genetically determined" (8).

For Lacan, this biological context determines the underlying "structural feature" (*Seminar I*, 209) of human culture. He states that "the human being is born with foetalised traits, that is to say deriving from premature birth" (210), so that the "little human being, like the little animal for a limited period, is stitched into the maternal companionship which saturates a certain primitive need he [sic] has.... But he is much more so than any other [animal], on account of the backwardness of his development" (209–10). Lacan uses this biological backdrop to indicate how the "foetalised" human infant develops into a member of human society via signification rather than via direct engagement with the "real world" ostensibly signified by this specifically human form of behavior. That is, nonhuman mammals are born with genetically determined instincts to avoid predators, access food and water, produce offspring, and so on. The human genome also possesses these instincts, but unlike other animals, humans develop their relationships to these instincts in a fundamentally communal, symbolic manner determined by our prematurity, and filtered through signifying matrices like language.

The repercussions of this unusualness are significant and can help to clarify the extent to which culture differs from other aspects of nature. These repercussions hinge on the ways that human culture overwrites the direct engagement that other animals have with their environments and that gets disrupted by humans' epigenetic development. This overwriting means that we do not have direct access to the Kantian "thing-in-itself." Instead, our access to this "thing" is filtered through signification. Lacan's use of the term "signification" is derived from the structural linguistics of Ferdinand de Saussure, which recognizes that the "linguistic sign unites, not a thing and a name, but a concept and a sound-image" (66). Signification involves two components, both of which are cultural overwritings of the Kantian "thing"—the signified is a concept, and the signifier is some way the signified is represented. But, crucially, neither component exists independently as an objective aspect of the external world. The English word "tree," and the drawing of a tree in Saussure's *Course in General Linguistics* (65), for example, are signifiers pointing toward a particular signified, but the concept which is signified is not an objective description of the "thing" existing outside humanity's phenomenological experience. Instead, the signified concept is entirely culturally constructed. And because, as Lacan puts it, "no signification can be sustained other than by reference to another signification" (*Écrits* 150), the concept signified by a signifier exists in relationship to other signifiers, not to objective material reality. Signification thereby alienates us from that objective material reality.

Signifiers link together to form what Lacan calls the "signifying chain" (*Écrits* 153). A return to a comparison with nonhuman animals' engagement with that which is external to them can help to clarify how this signifying chain relates to our own engagement with the external. Nonhuman animals live in environments comprising sources of nutrition, dangers like predators, other members of the same species who may be rivals or mates, and so on. Since, as Darwin argues, every organism, "has to struggle for life, and to suffer great destruction" (61), the animals which are able to pass on their genes through sexual reproduction are those which are able to successfully respond to these environmental stimuli, and in so doing the descendants of those animals are genetically predisposed to also successfully respond to those environmental stimuli. There is something like a direct one-to-one mapping here between the animal and the environment, so that this last term conforms to both elements of the meaning which is typically ascribed to it—the nonhuman animal's environment is both an "environs," that which is around the animal, and also entirely natural in the sense that the animal responds to environmental stimuli according to genetic dictates. Existing ecocriticism often uses Dawkins' term "cultural environment" (193) to describe human engagement with the external world. The "environment" component of the term "cultural environment" makes a similar suggestion that this culture is not only an "environs" around humans, but also entirely natural, in the sense that human responses to cultural environmental stimuli also operate according to genetic dictates. In fact, to complicate matters, those utilizing the concept "cultural environment" (and we could position the use of memes in relation to adaptation in this context), are not generally explicit about whether human responses to cultural environmental stimuli are precisely determined by genetic dictates or not. It is perhaps more the case that the term "cultural environment" implies this natural/genetic component rather than states this directly. Either way, the term "signifying chain" is more accurate than "cultural environment" in explaining how humans engage with the world, because it emphasizes how that engagement is predicated on epigenetic as well as genetic factors that prevent us from responding to external stimuli at the exclusively genetic level, and because it provides a precise account of how those epigenetic factors take shape—at the alienated, symbolic level of signification. The paradoxically naturally evolved unnaturalness of signification prevents our direct engagement with the external, so that we see, as it were, through a glass darkly, alienated in what Lacan calls the symbolic order of concepts, rather than directly present in an environment of the things referred to by those concepts.

Textual adaptation operates within this broader context of human culture as the epigenetic result of genetic predicates. The rest of this essay is devoted to addressing the repercussions of conceptualizing adaptation according to this psychoanalytic model versus conceptualizing adaptation according to the logic of memetic evolution. For the memetic model, certain stories have some kind of thematic importance for a particular culture. Cultural conditions are not static, so the story evolves to fit into new cultural environments. There are two main forms of active agent in relation to these changes—changing cultural conditions and authors/audiences who adapt stories to fit into these changing conditions. The first form of active agent is the changing cultural conditions, which are themselves overdetermined by people, but also by material contingencies, including explicitly noncultural contingencies like droughts, volcanic eruptions, earthquakes. The example of "Noah" that I'm coming to provides a clear example of such noncultural contingencies having a cultural influence because the hyperbolic quality of the divine deluge exaggerates a natural, material phenomenon, as Clare Amos notes:

> The biblical flood cannot be understood apart from the Mesopotamian accounts of the flood [principally the *Epic of Gilgamesh* and *Atra-Ḫasīs*] to which it owes several significant features. . . . The geography and climate of Mesopotamia encouraged reflection of a "flood": civilization in Mesopotamia, the land between the great Tigris and Euphrates rivers, was only possible because of the annual flood of the rivers [without which] the elaborate irrigation system of the Mesopotamian cities would dry up:

> yet in years when there was an excess of water destruction and death would result. Water brought both death and life. (67–8)

Noncultural factors in the material environments in which cultures are situated, then, have an impact on those cultures and on the stories they tell.

The second group of active agents in the meme model are authors and audiences who adapt the story to fit into different cultural environments. The extent to which this group has genuine agency is open to debate, given the way that Dawkins, first proponent of memes, states that "when you plant a fertile meme in my mind you literally parasitize my brain, turning it into a vehicle for the meme's propagation in just the same way that a virus may parasitize the genetic mechanism for the host cell" (192). If Dawkins' argument is followed to the letter, then it is not entirely clear that there is also a nonparasitized component to the human brain/culture over which humans have direct autonomy. However, the meme model tends to be used within adaptation studies in a less deterministic manner. When Linda Hutcheon proposed this approach, she argued that "evolving by cultural selection, traveling stories adapt to local cultures, just as populations of organisms adapt to local environments. We retell... stories over and over; in the process, they change with each repetition, and yet they are recognizably the same" (177). Here, "cultural selection" has a similar environmental function to natural selection, providing the context in which texts, like organisms, might either survive or perish, and those texts themselves "adapt" because "we retell" them. This is a less deterministic and less parasitic process for Hutcheon than for Dawkins.

The signification model is no different from the meme model in agreeing that stories have some kind of thematic importance for particular cultures. The signification model also agrees that cultural conditions are not static. However, a divergence begins at this point. For the meme model, stories evolve to fit into changing cultural environments. In contrast, for the signification model, stories do not behave like the exclusively natural phenomenon of evolution, nor are the changing cultural contexts in which stories get retold exclusively natural. To use the same example, the Mesopotamian precursors to "Noah" in the *Epic of*

Gilgamesh and *Atra-Ḫasīs* are produced by a culture which had important engagements with the phenomenon of rivers flooding. However, members of that culture did not experience flooding only as an exclusively natural phenomenon; instead, symbolic mediation meant that flood waters were experienced as signifiers and signifieds relating to concepts like benevolence, protection, order, guilt, punishment, mercy, virtue, and so on. Stories are one of the cultural phenomena that make sense of alienated natural phenomena through symbolic signification. The cultural milieu is therefore not an environment, in both senses of the word, but a signifying chain which structures how that culture makes sense of the external. And when cultural conditions change, either as a result of the signifying chain having its own transformative momentum, or from more definitively external pressures like floods or changes in climate, the ways that signifiers relate to one another also change. The signifying chain is reshaped, and stories are retold to fit within the restructured signifying chain.

As with the meme model, there are two main forms of active agent in relation to these changes. The first is the same as in the meme model—changing cultural conditions. The second agent is different, though—the signifying chain itself. Signification determines how humans respond to stimuli, whether those be explicitly environmental stimuli like floods, or explicitly cultural stimuli like existing forms of signification. Malcolm Bowie demonstrates the passivity of human subjects in relation to signification by adapting Dawkins' argument that genes are the "fundamental units of natural selection, the basic things that survive or fail to survive" and that these genes generally "gang together into large communal survival machines or 'vehicles'" (Dawkins 253) which are organisms. Dawkins' "selfish gene" thus uses the unwitting organism to reproduce itself. Bowie uses a similar biological analogy to explain the agency of signification when he states that "just as the hen is an egg's way of producing another egg, so the characteristic sensations of 'being a person'... come from the self-perpetuating imperative that propels the signifying chain" (76). Humans, then, are the survival machines or "vehicles" that the signifying chain uses to create more signification. This, in fact, is a more accurate analogy between biology and culture than that proposed by the meme model—in biology, organisms are replicated only so that the active agent, the selfish gene, may replicate itself; similarly, stories are replicated only so that the active agent, the (selfish) signifying chain, may replicate itself. The difference between Dawkins' and Lacan's argument, however, is the fact that signification is not an entirely natural phenomenon, given the relative autonomy of signification in relation to the typical dictates of genetic evolution. Signification is relatively autonomous from all other products of genetic evolution because even though evolution generated signification, in a roundabout way via humanity's genetic prematurity, once generated that signification does not correspond to any other aspect of nature, and in fact brings about an alienation from direct engagement with the rest of nature which is itself agentic and determining of human behavior. This last claim, a central tenet of Lacanian psychoanalysis, is counterintuitive and even offensive to the ostensible plenitude of the rational Cartesian ego. We experience ourselves as active agents; consider ourselves capable of autonomous choice and possessors of free will. But, for Lacan, these experiences are effects generated *by* signification rather than causes *of* signification. It is not so much we who speak words as words which speak through us. As Lacan puts it, the human "subject is nothing other than what slides in a chain of signifiers, whether he [sic] knows which signifier he is the effect of or not. That effect—the subject—is the intermediary effect between what characterizes a signifier and another signifier" (*Seminar XX*,50). We are the effect of signifiers—our response to a natural phenomenon like a flood is not something we determine ourselves on the basis of objective rationality, nor something solely determined by evolved instincts to protect the genes within us or our loved ones, nor even determined only by a memetic exaggeration of our instinct to protect shared genes which might take the form of rescuing valued heirlooms and so on. No, our response to a phenomenon like a flood is principally determined by our positionality within signifying chains composed of the aforementioned signifiers and signifieds relating to concepts like benevolence, protection, order, guilt, punishment, mercy, virtue, and so on.

Stories are therefore subsidiary components of any given signifying chain. Yes, the signifying chain has some relationship to material conditions, but it also provides abstract, alienated conceptualizations of those conditions. So a culture that produces food from irrigated rivers which occasionally cause devastating floods conceptualizes these environmental conditions through a signifying chain that is composed of subsidiary components like the flood myths in the *Epic of Gilgamesh* and *Atra-Ḫasīs*. Cultures change both through the vagaries of environmental conditions and through the relative autonomy of signification, and stories get adapted by the agentic signifying chain to fit into newly arranged patterns of signification, and to become subsidiary components of those signifying chains.

The "Noah" Story in Relation to Cultural Environments and Signifying Chains

The intertextual afterlife of "Noah" includes aspects of continuity and aspects of change from the hypotext that can be explained by both the meme model and the signification model. Identifying the salient components within "Noah" that will subsequently be adapted is complicated by the fact, mentioned above, that the biblical text is itself adapted from earlier sources. Nevertheless, components that will subsequently relate to ecologically relevant retellings can be noted, as follows. The biblical flood is not only divinely ordained but is so ordained because of humanity's wickedness. God rewards the only person He deems virtuous, Noah, by ordering him to build an ark, so that his family and two of every animal will be able to survive. The flood does indeed wipe out the rest of humanity, but after forty days and nights the inundation ceases, and the ark makes landfall on a mountaintop. Noah sends out birds to check that the floods have subsided and then sacrifices to God in celebration, with a divine promise in return that God will not again flood the Earth, which should be replenished by Noah's family.

Recent adaptations of the "Noah" story keep the narrative event of a small number of characters surviving an apocalyptic deluge aboard an ark-like vessel, but change the causes of the deluge. Films such as Roland Emmerich's *2012*, for example, depict the events where most of the world's population are killed as raging tsunamis cover the globe, or are set in the aftermath of global flooding, like Kevin Reynolds' *Waterworld*. Either way, small groups of characters will survive these events aboard ark-like boats, which are explicitly given that title in *2012* (where the main protagonist's son is also named Noah).

The main difference between these retellings and the biblical hypotext is the cause of the disaster. As Amanda Norsker notes, the "importance of Noah's righteousness has its basis in Yahweh's motive for bringing the flood: the humans have become evil and the earth is corrupted" (56). Contemporary retellings remove this supernatural motive. Instead, the disasters have natural causes, such as a solar flare that causes the Earth's core to overheat and destabilize the planet's crust, in the case of *2012*, or melting polar ice caps in the case of *Waterworld*. However, the hypertexts do keep the notion of these events being punishment for evil, and those who survive being virtuous, so that some aspects of the source are altered and some are retained. It is also the case, though, that even the component which is retained—humans either survive or are punished because they are either virtuous or evil—undergoes an important transformation, in terms of what makes humans evil. The biblical account is vague about what it is that makes the pre-deluge humans evil, with the only specific vice mentioned (twice) being violence (*King James Bible*, Genesis 6.11, 13). What is also mentioned twice at this stage, though, is that God created "man" (6.6–7), so that humanity's evil is juxtaposed with the good which God has mandated. Hypertextual films will make an important change in this regard, because although they still imply that those who are vaguely responsible for the disaster are violent, this implication has no direct connection to a "correct" form of behavior that God has ordained. Rather, these films depict humanity's evil as a departure from a "natural" rather than a divine goodness. *Waterworld*'s antagonists the Smokers, for example, who are trying to steal information about how to reach the nonflooded mountaintop called "Dryland" from the film's protagonists, have their base on the rusty hulk of the *Exxon Valdez* and use automated vehicles that pollute the air and their bodies with dark smoke and oil, in contrast to the film's hero, the Mariner (Kevin Costner), who uses more ostensibly "natural" wind power to propel his sailboat. *2012* will punish the similarly polluting behavior of oligarch Yuri (Zlatko Burić), who tries to escape the disaster aboard his ostentatiously large aircraft filled with sportscars. Even Darren Aronofsky's more explicit adaptation of the biblical source,

Noah, which differs from the aforementioned films in the fact that the disaster *is* caused by God, emphasizes the ecologically damaging component of humanity's evil, with the destroyed antagonists shown covering the Earth in cancerous and polluting industry, cutting down forests, and eating meat in contrast to Noah's (Russell Crowe) vegetarianism. And, if these films keep the notion of the wicked being punished, but change the reasons *why* humanity is wicked, then they also keep the notion of the virtuous being rewarded, but change the reasons why the protagonists are virtuous. In fact, it is more accurate to say that the contemporary hypertexts sidestep the virtuousness of the protagonists—it is sufficient that they can be juxtaposed with the more definitively "unnatural" corruption of other characters. In *2012*, for example, Jackson's heteronormative family is initially fractured by a usurping stepfather, a plastic surgeon who performed breast enlargement surgery on Yuri's girlfriend Tamara (Beatrice Rosen). These last three characters will be judged sufficiently unnatural to be punished with death, and Jackson's family sufficiently virtuous in contrast to be reunited. Both the biblical and contemporary texts not only allow certain characters to survive, then, but renew and "cleanse" those characters, even if the two cultures which produce these texts conceptualize evil and virtue very differently.

Before addressing the ways that the evolutionary and psychoanalytic methods precisely conceptualize these consistencies and changes, it is possible to summarize the following: The hypotext and hypertexts demonstrate that biblical culture and contemporary culture both have anxiety about humanity's perceived evil/corruption, and try to make sense of that anxiety through narratives that punish the guilty and reward the virtuous via a deadly flood. The changes in the texts demonstrate, however, that the two cultures conceptualize that evil/corruption differently, with the biblical culture framing evil/corruption as an alienation from the divine, and with contemporary culture framing evil/corruption as an alienation from "nature."

Both the evolutionary and the psychoanalytic approaches can account for these aspects of adaptational continuity and change, but there is a decided advantage to the latter approach. If the intertextual afterlife of "Noah" is conceptualized according to the logic of memetic evolution, then continuity and change relate to different cultural environments.

Comparing biblical and contemporary cultures, certain features of those cultures are different (the earlier culture frames corruption and virtue in relation to God; the later culture frames corruption and virtue in relation to "nature"), but certain features of those cultures are similar (some humans are corrupt and some are virtuous, and both cultures make sense of this by imagining quasi-apocalyptic punishment and reward for this corruption/virtue). Changes or consistencies between hypotext and hypertext indicate that the two different cultural environments have certain similarities and certain differences.

If the intertextual afterlife of "Noah" is conceptualized using the psychoanalytic model, then continuity and change relate to different structures of signification. This model agrees with the evolutionary approach that certain features of the two cultures are different and certain features are similar. However, changes or consistencies between hypotext and hypertext do not indicate something so ostensibly "natural" as two different cultural environments having certain similarities and certain differences. Rather, those similarities and differences operate within the alienated context of the symbolic order. The configurations in which signifiers are chained together are different for each of those cultures. The intertextual afterlife of "Noah" demonstrates how a story can be reiterated to facilitate signification in a new cultural context—if feature *x* of a story no longer carries important signifying weight (God no longer monopolizes the position of moral judge or existential threat), then that feature can be changed because feature *y* of a story (corruption; punishment; virtue; survival; cleansing) continues to signify. Stories about devastating floods are therefore reiterated according to the structure of the new symbolic order—the active agent is neither the story itself (evolving like an organism to fit into a changed environment) nor an author who consciously repurposes the story by altering feature*x*, which once fit into a different cultural context, but which is now obsolete. Instead, it is the signifying chain itself that is agentic and determining. It provides the alienated, abstract coordinates that appear to directly and objectively map onto that which is signified. We are not in control of ideas like corruption, punishment, virtue, survival, and cleansing. These ideas are, to use Lacan's neologism, extimate—simultaneously generated by external signification but experienced intimately as though autonomously generated from within. Thus, the story and the author act under the pressure of the internalized symbolic order, which necessitates that the story be restructured to conform with the broader signifying chain.

Conclusion

The psychoanalytic approach not only conveys a more accurate picture than the evolutionary approach but also, in emphasizing the agency of the signifying chain, demonstrates how "making sense" is the predominant concern of that agent. The meme model might seem more optimistic, with a suggestion that the cultural environment has shifted in a manner which has become aware that humanity is at an ecological precipice, and we therefore tell stories about that crisis to motivate real-world action. Things are not so simple for the psychoanalytic approach, however. Yes, as mentioned above, cultures change due to cultural and noncultural factors. The climate crisis is a clear example of how such cultural and noncultural factors intersect, and our culture is clearly changing under its pressures, with the signifying chain adapting itself in the process. But contemporary retellings of "Noah" demonstrate that at least one element of the signifying chain has not changed in an important and damaging regard—anxiety about evil/corruption continues to be resolved at the fictional level, with disasters punishing the wicked and sparing the righteous, even if the reasons why characters are deemed wicked or righteous has changed. The intertextual afterlife of "Noah" therefore conforms to Fredric Jameson's argument that even when stories address issues that challenge the structure of the symbolic order, they are to be "grasped as the imaginary resolution of a real contradiction" (77). The signifying chain is more concerned with managing contradiction/crisis than with resolving contradiction/crisis, and with reproducing itself than with the wellbeing of the human hosts it uses to reproduce itself. The crucial ideological aspect of "Noah" that contemporary adaptations retain is this imaginary resolution of crisis, with its promise that all will get what they deserve.

The psychoanalytic method is not only more accurate than the evolutionary approach, in terms of identifying precisely how cultural activity like textual adaptation differs from other aspects of nature. The psychoanalytic approach is also more useful in the sense that it identifies exactly how textual adaptation can reproduce illusory resolutions to real problems. Given that at least one key task of ecocriticism is the urgent need to explain why our political, social, and cultural institutions are not making the necessary changes to avoid and ameliorate real-world ecological disaster, the psychoanalytic account identifies one enduring aspect of that self-destructive passivity.

References

2012. Directed by Roland Emmerich, Columbia, 2009.
Amos, Clare. *Birthpangs and Blessings*. Sacristy Press, 2004.
Ansermet, François and Pierre Magistretti. *Biology of Freedom*. Karnac Books, 2007.
The Bible. King James Version, Oxford UP, 1998.
Bortolotti, Gary, and Linda Hutcheon. "On the Origin of Adaptations." *New Literary History*, vol. 38, no. 3, 2007, pp.443–58.
Bowie, Malcolm. *Lacan*. Fontana, 1993.
Darwin, Charles. *On The Origin of Species*. John Murray, 1859.
Dawkins, Richard. *The Selfish Gene*. Oxford UP, 1976.
Elliott, Tomas. "'A Movie about Flowers?' Notes on the Ecological Turn in Adaptation Studies." *Adaptation*, vol. 17, no. 2, 2024, pp. 320–37.
Gould, Stephen Jay. *Ontogeny and Phylogeny*. Harvard UP, 1977.
Hutcheon, Linda. *A Theory of Adaptation*. Routledge, 2012.
Jameson, Fredric. *The Political Unconscious*. Methuen, 1981.
Johnston, Adrian. "Drive Between Brain and Subject." *The Southern Journal of Philosophy*, vol. 51, 2013, pp. 48–84.
Lacan, Jacques. *Écrits: A Selection*. Translated by Alan Sheridan, Tavistock Publications, 1977.
Lacan, Jacques. *The Seminar of Jacques Lacan: Book I*. Translated by John Forrester, Cambridge UP, 1988.
Lacan, Jacques. *The Seminar of Jacques Lacan: Book XX*. Translated by Bruce Fink, Norton, 1998.

Malabou, Catherine. "The King's Two (Biopolitical) Bodies." *Representations*, vol. 1, 2014, pp. 98–106.

Noah. Directed by Darren Aronofsky, Paramount, 2014.

Norsker, Amanda. "Genesis 6,5–9,17: A Rewritten Babylonian Flood Myth." *Scandinavian Journal of the Old Testament*, vol. 29, no. 1, 2015, pp. 55–62.

Saussure, Ferdinand de. *Course in General Linguistics*. Translated by Wade Baskin, Columbia UP, 2011.

Waterworld. Directed by Kevin Reynolds, Universal, 1995.

White, Jr., Lynn. "The Historical Roots of Our Ecologic Crisis." *Science*, vol. 155, no. 3767, 1967, pp. 1203–7.

3

A History of Urgency: Cultural Transformations and Ecological Reflections in Czech Environmental Documentaries

Petr Bubeníček and Tereza Dědinová

This case study offers a historical perspective on documentary films reflecting the environmental crises in Communist Czechoslovakia and the Czech Republic from 1978 to the present. The study draws from adaptation studies, intermedial studies, and ecocriticism to examine how documentaries have contributed to the public discourse on environmental protection. Integrating ecocriticism with intermedial and adaptation studies provides cognitive frameworks for understanding how documentaries engage with and respond to ecological crises.

In discussing selected documentaries "adapting" the ecological crisis—covering environmental issues, intellectual debates, scientific reports, historical and spoken testimonies about environmental movements, and

P. Bubeníček (✉) · T. Dědinová
Department of Czech Literature, Faculty of Arts of the Masaryk University, Brno, Czech Republic
e-mail: petrb@phil.muni.cz

T. Dědinová
e-mail: terezadedinova@phil.muni.cz

P. Demory (ed.), *Ecoadaptation*, Palgrave Studies in Adaptation and Visual Culture,
https://doi.org/10.1007/978-3-032-09847-4_3

active environmental protection—we will examine how these films interpret the ecological problems of their time and negotiate environmental meaning through their narrative, aesthetic, and affective strategies. The scope of this study does not allow us to explore the diversity of ecologically oriented documentary filmmaking comprehensively. Rather, it invites readers to explore the possibility of taking adaptation and intermedial approaches to environmental history. As historian Christof Mauch argues, incorporating nature as an active agent in historical narratives challenges conventional models of causality, disrupting linear assumptions about human intention and outcome: "When we include nature as an important player in our historical narratives, our established models of thinking, which assume a direct causal connection between intentions and their realization, suddenly cease to function as smoothly as we have been accustomed to" (15). This perspective is particularly relevant when analyzing documentaries that reflect shifting environmental concerns and ideological contexts. To explore this dynamic, we will focus on three films that form a historical arc from state socialism—referring to the political and economic system of communist Czechoslovakia—to today's neoliberal consumer capitalism: Václav Hapl's *Lesy umírají tiše* (1978), Jan Svěrák's *Ropáci* (1988), and Barbara Klocová's *Lesní žal* (2023).

Defining documentary film poses a challenge, especially as this genre constantly evolves and encompasses diverse styles and approaches. This difficulty in defining documentary film is captured by Jonathan Kahana, who characterizes it as "a notoriously slippery eel, perhaps the oldest and slipperiest concept in the history of cinema's public and commercial modes and genres" (1). Any innovative, experimental, or mockumentary approach by a documentary filmmaker challenges conventional definitions of the genre, pushing us to rethink its relationship to reality, modes of representation, and audience engagement. Nichols describes documentary film as a medium that conveys actual events and invites viewers to engage with real people and situations, shaped by the filmmaker's perspective: it offers "a plausible proposal about, or perspective on, the lives, situations, and events portrayed," allowing us to see "the historical world directly rather than... a fictional allegory" (14). Documentary filmmakers can thus be understood as both observers and interpreters.

Documentary film provides a factual account of the world and, above all, creates a perspective through which we view its complexity.

Ecocriticism, as a broad field of literary and cultural studies, brings to the humanities a focus on the world beyond the text. According to the widely cited definition, proposed by Cheryll Glotfelty, it is "the study of the relationship between literature and the physical environment" (xviii). Growing out of Joseph Meeker's notion of literary ecology introduced in *The Comedy of Survival: Studies in Literary Ecology* (1974) and literary scholars' unease with the escalating environmental crisis, ecocriticism has evolved from its humble beginnings in American studies to a global phenomenon that examines cultural texts of diverse genres and origins (Clark 137).[1] Today, ecocriticism is a profoundly interdisciplinary field within the environmental humanities that bridges literary studies, ecology, philosophy, cultural studies, history, gender studies, visual arts, environmental communication, and political science to critically examine the complex relationships between humans, culture, and the environment. In this chapter, we will examine the selected documentaries from the ecocritical perspective of transcorporeality, which focuses on the intermeshing of bodies with the material world.

[1] Since the first wave of ecocriticism's initial focus on nature writing and the celebration of wilderness, the field has expanded to explore a wide range of relationships between culture and nature, including those in urban areas, as well as environmental justice issues. Ecocritical scholars examine the concept of nature and its representations in the arts, and also place a growing emphasis on the theoretical foundations of the discipline. Beyond its engagement with environmental literature, ecocriticism now extends to "the scrutiny of ecological implications and human–nature relationships in any literary text (or other artistic text), even texts that seem, at first glance, oblivious of the nonhuman world" (Slovic, "Ecocriticism" 4).

From Exploitation to Awareness: The Shifting Ecology of Czechoslovakia and the Czech Republic

Forty years of Communist rule devastated the environment in Czechoslovakia. Communist ideology emphasized man's control over nature and the ability of a centrally planned economy to alleviate environmental problems through technological progress. From its very establishment, the economy, and indeed society, in Communist Czechoslovakia, was based on heavy industry and was unscrupulous to the biosphere. Following the Stalinist model, the political elites sought to exploit nature without considering its ill effects. Radical transformations in agriculture also contributed to environmental degradation. In the wake of collectivization—a process politically enforced under state socialist regimes to eliminate private farming—land was merged to create vast monocultural fields, resulting in long-term ecological imbalances due to excessive fertilization, pesticide application, and large-scale land drainage, the consequences of which remain evident today. The situation peaked in the 1970s and 1980s, the period of so-called normalization (which refers to the period in Czechoslovakia after 1968 when Communist control was reasserted, suppressing dissent and enforcing conformity until 1989), when the air was fouled with pollution, toxic waste poisoned the land and waters, and acid rain destroyed the forests. The devastation reached catastrophic proportions in northwestern Bohemia, where the chemical industry was concentrated, and brown coal was extracted from opencast mines and burned in emissions-spewing power stations.

The centralized, authoritarian Communist regime censored information to maintain the government's positive image. Given the regime's nature, it is unsurprising that the state concealed, suppressed, or downplayed negative environmental facts to prevent citizens from learning about the problems. Such behavior maintained the appearance of political legitimacy and control. In the late 1980s, the unsustainable ecological conditions and the alarming impact of industrial pollution on human health spurred previously unengaged citizens into action, and environmental protests intertwined with protests against the totalitarian establishment (e.g., Vaněk).

During the harshly repressive 1950s, the germs of a new way of thinking emerged, one that questioned the unbridled exploitation of natural resources and attempted to redefine the relationship between humans and the environment. Zoologist Otakar Leiský played a crucial role when in 1958 he initiated the creation of the Nature Conservation Section of the Society of the National Museum, which in the late 1960s transformed into the independent *Tis: Union for Nature and Landscape Conservation Organization*, combining the romantic forest wisdom of Ernest Thompson Seton with biological expertise (Jehlička and Smith 84). *Tis* was devoted to conservation work, lectures, and youth education (86–87) but faced pressure during normalization, culminating in its dissolution in 1979, when it was replaced by the *Czech Union of Environmentalists* and the *Brontosaurus* youth movement, which offered some freedom (engaging in environmental activities, educational programs, organizing conservation projects) to those interested in nature conservation despite state control. Somewhat unexpectedly, the international *Ecofilm festival* was established in the mid-1970s as a platform for discussing environmental protection and global perspectives. This paradox reflects the regime's contradictory stance. While suppressing criticism, it simultaneously tolerated certain platforms engaging with ecological issues, perhaps under the guise of scientific or educational dialogue.

In the 1980s, ecological issues became something of a "tolerated zone," as they did not initially appear overtly political. This, in turn, contributed to the possibility of producing the student film *Oil Gobblers*, which will be discussed later in this study. The *Ecological Section of the Czechoslovak Biological Society* provided a platform for interdisciplinary debate, and many of its members would later become key figures in the ecological turn following the Velvet Revolution of 1989. By the late 1980s, ecological activism had increased through petitions, open letters, and demonstrations, connecting environmental concerns with broader social issues (Vaněk 154). These developments reflect the gradual emergence of environmental awareness during state socialism and foreshadow the significant shift after 1989, when environmental policy was increasingly displaced by market pragmatism.

The Poetic Mode: Forests Die Silently

In *Introduction to Documentary* Bill Nichols explores what he dubs the poetic mode of documentary film. Documentaries in this mode prioritize an aesthetic dimension, draw on the traditions of avant-garde cinema, and frequently incorporate visual metaphors (116–121). The poetic mode is exemplified in the work of Czech filmmaker Václav Hapl (1928–1982).[2] His fourteen-minute 1978 documentary *Lesy umírají tiše* (*Forests Die Silently*) addresses the themes of forest fires and conservation. Yet the film does not function as a straightforward instructional piece. Instead, it prioritizes visual beauty, rhythm, and atmosphere, offering an expressive treatment of its subject.

Hapl's expressive use of camera movement, editing, and sound is part of his artistic intent. The camera alternates between slow movements and fast sequences. One section contrasts the camera's calm upward movement, capturing trees, with rapid horizontal movements depicting their destruction. Elsewhere, a fast, dramatic montage jarringly contrasts with calm shots of the untouched forest. Hapl further challenges traditional documentary conventions with staged sequences: for example, scenes depicting a family getting out of a car or a man starting a fire with a cigarette butt. Fire plays a central role in the documentary. Repeated footage of individuals lighting fires despite bans, or discarding cigarette butts, and the resulting devastation illustrates the direct consequences of human actions on forests.

[2] After 1948, Václav Hapl worked at the Prague studio of Krátký film Praha (Short Film Prague), and from 1952 at the Czechoslovak Army Film (Československý armádní film), where he remained until 1970, producing a number of instructional films. His position became particularly difficult in the early 1970s, when he was dismissed from the Czechoslovak Army Film following political vetting. In 1972, he returned to Krátký film, where he directed documentaries that moved toward the essay film form.

The voice-over, characteristic of contemporary ecological documentaries, depicts forests as the last islands of wilderness in regions once covered by vast woods, memorials to a time when "humans and nature understood each other." The narration contrasts the past, when ancient hunter-gatherers lived in and off the forests, with the present, when people visit the dwindling forests like a museum. Likening a forest living ecosystem to a museum is itself problematic. Moreover, while this comparison suggests respectful attention of visitors toward the exhibits, the film portrays people seeking recreation in the forests as inconsiderate of the environment and as contributors to the ongoing ecological crisis. Their sheer numbers pose the problem. Instead of solitary pilgrims "who honor the order of nature," families with cars, pets, and camping gear flood forests, using them for relaxation and leaving litter, trampled soil, and poorly secured fires that risk catastrophic destruction.

The forests are portrayed as possessing agency, with active verbs highlighting their functions: "They give wood, protect from erosion, make the air breathable." A reflective question follows: "Forests give their wealth to everyone. Do we give back to the forests in the same way?" A direct appeal underscores human responsibility in the fate of forests: "Do not blame nature; lightning rarely causes forest fires." Humans are thus depicted as a primary threat to wild forests, increasingly replaced by artificial plantations for wood production.

Hapl points to the strained relationship between industry and nature, and criticizes industrial air pollution as a major cause of forest loss. Impressive shots pair fume-belching chimneys, mining wastelands, and dead, withered trees, with a somber commentary, "Thousands of trees are dying in the polluted air, and future generations will miss them." Yet instead of offering a broader critique of industry and the political system, the film emphasizes individual responsibility for protecting forests. This personal appeal distracts from the need for systemic change.

Human Bodies in the Forest/World

The concept of **transcorporeality**, developed by Stacy Alaimo ("Bodily") to emphasize the interconnectedness of human and animal bodies with the material world, appears to be a particularly inspiring framework for analyzing the relationships of individuals with nature and proves useful in understanding the ecological impact of Hapl's film. Alaimo claims that the human body is not merely a biological entity but is also a political, social, ecological, and cultural subject situated at the intersection of environmental flows of materials and substances, energies and processes ("States"). This framework undermines traditional Western philosophical and cultural notions of human separateness from and superiority over nature, instead highlighting humans' physical and ethical entanglement with the broader world.

Transcorporeality, therefore, rejects the idea of nature as something "external" to humanity. Alaimo demonstrates how environmental substances—often toxic (e.g., pollutants, microplastics)—permeate our bodies, dissolving the boundaries between nature and culture. Within the framework of critical posthumanism, which rethinks ecological, cultural, and ethical relationships among humans, nonhumans and technologies, transcorporeality redefines the subject as a "transversal entity"—a term Alaimo uses to highlight the interconnected nature of existence, where boundaries between the human and nonhuman, the material and discursive realms are fluid, encompassing not only humans but also animals and the planet as a whole (Alaimo, "Trans-corporeality"). The ethical dimension of transcorporeality stresses that everyday human actions have far-reaching consequences for both human and nonhuman lives (Alaimo, "States"). However, this is not the romanticized notion that "everything is connected to everything else"; rather, it is a radical call to rethink our ontologies and epistemologies.

In this chapter, we adapt the concept of transcorporeality for the analysis of documentary films by redirecting its focus away from how the environment "enters" the body. Instead, we shift our attention to how human bodies—their needs, emotions and activities—shape and transform the environment, and, in turn, how these bodies and minds are reciprocally influenced by the environment as it undergoes transformation. Beyond the physical aspect, we emphasize the emotional dimension of human–environment interactions, particularly in natural and forested landscapes. This approach, in line with ecophenomenology developed by David Abram (e.g., in his seminal study *The Spell of the Sensuous: Perception and Language in a More-Than-Human World*, 1996), focuses on embodied, sensory, and emotional experiences of the natural world.

From a transcorporeal perspective, the language used to narrate Hapl's film is noteworthy: it bridges the gap between culture and nature by employing natural metaphors to describe humanity's impact on forests. For example, clear-cutting is likened to the aftermath of a hurricane, human-caused fires to avalanches. In the visuals and the narration, humans and forests are closely connected, yet their relationship is portrayed as fundamentally unbalanced: the forest gives, while humans destroy. The presence and demands of humans threaten the forest; even their desire to spend time in nature proves harmful. Human hands discard cigarette butts, only to later try in vain to extinguish the resulting raging fire. Human corporeality and the need to be in contact with nature, when coupled with a lack of respect for the nonhuman environment, pose a significant threat to the survival of forests—ecosystems vital for human life.

Unlike Barbara Klocová's documentary, which we analyze below, Hapl does not address the emotional toll of environmental degradation on humans. However, *Forests Die Silently* prompts a broader ethical reflection: What aspects of our world—natural, cultural, or social—do we consider worthy of protection, and what steps are we willing to take to ensure their survival? The film rejects inevitability, the idea that the forces of nature or history make decisions without human input. Instead, it focuses on human activity, reflection, and responsibility. Although the documentary avoids political commentary, it highlights the problem of perceiving nature primarily as a resource—here, a source of wood and recreation—rather than a complex ecosystem.

The film presents forests not just as ecological entities; they symbolize a deeper relationship between humans and nature. Their survival depends on the collective decision to value them as essential to life, beauty, and ecological balance. By embracing the poetic mode, Hapl invites viewers to see forests as part of a larger existential inquiry: What kind of world do we want to create and sustain for future generations? The film thus becomes a call to action, urging us to prioritize the survival of forests—not because they will survive regardless of our actions but because we choose to ensure their survival as part of a vision of a sustainable future.

Ecology Upside Down: Oil Gobblers

Bill Nichols defines mockumentaries as cultural expressions that "adopt documentary conventions but are staged, scripted, and acted to create the appearance of a genuine documentary and leave clues that they are not. Part of the pleasure they provide lies in how they let a knowledgeable audience in on the joke: we can enjoy the film as a parody and gain new insight into taken-for-granted conventions" (12). This definition applies to Jan Svěrák's 1988 film *Oil Gobblers*, created as a graduation project at the Film and TV School of the Academy of Performing Arts in Prague. The school's relatively open environment facilitated the production of a film that has since become a cult classic within the Czech environmental movement. The "joke" Nichols refers to lies in the film's premise: it follows a scientific expedition searching for an animal that does not exist, parodying the conventions of scientific and exploration documentaries.

A fictional scientific expedition travels to one of the most ecologically devastated areas of the former Czechoslovakia, northwestern Bohemia. This landscape, scarred by large-scale lignite mining and emissions from factories and coal-burning power plants, is where a new fictional species—the "oil gobbler"—has evolved. This creature thrives in inhospitable conditions, eating plastic, drinking oil, and breathing polluted air. Although the implausibility of this narrative is gradually revealed, the film strives to maintain a formal impression of verisimilitude. The scientists behave like they are part of a real expedition and carefully document their findings. Every detail is presented as if it were a credible report by naturalists zealously pursuing their scientific goal: the discovery and confirmation of the creature's existence.

The mockumentary subverts the concept of environmental stewardship, which typically refers to the responsible use and protection of the natural environment. In *Oil Gobblers*, the scientific characters express unvarnished joy at the idea that the environmental degradation of Bohemian industrial areas will improve the living conditions of oil gobblers, envisioning that within a few years these creatures would thrive across large parts of the country. Their focus on the fictional animal's "improving situation" is entirely disconnected from the well-being of all other species, including humans.

Exposure to toxic waste, polluted water, coal-seam fires, and choking smoke is unhealthy—if not outright dangerous—for the human characters; the expedition's cameraman loses consciousness after being lowered into a steaming mine and must be rescued. Humans and oil gobblers, like all species, are deeply dependent on their environment; their corporeality renders them vulnerable to environmental changes. However, while the vulnerability of oil gobblers is acknowledged, and the human characters show deep concern for their welfare, the fate of other human and nonhuman animals is glaringly absent from the narrative. The film is conspicuous by its lack of direct commentary on the deplorable state of the landscape or criticism of industrial devastation—the consequences of prioritizing mining and industry over other, not solely human, needs for a healthy environment. The transversality of the human body (see Alaimo above) is expressed by the vulnerability of human characters in a manmade, inhospitable environment. However, this is alienated and unreflected within the parody mode. The film ends on a darkly ironic note: the scientists express hopeful certainty that the oil gobblers will never appear on the IUCN Red List of Threatened Species—a prospect that if achieved would come at the unspoken cost of many other species likely ending up on that very list.[3]

[3] For a more detailed analysis of *Oil Gobblers* and its environmental and cultural significance, see Bubeníček, "Irony and Catastrophe."

The Grief of the Forest: Youth, Solastalgia, and the Struggle for Renewal

Although the early 1990s in Czechoslovakia were marked by hope for environmental progress, environmental activists in the Czech Republic view the subsequent decades as problematic. While pollution levels decreased, environmental degradation continued due to industrial agriculture, the prioritization of economic interests over ecological concerns, and insufficient climate change action. Activists argue that these trends have resulted in missed opportunities, such as a faster transition to renewables or biodiversity protection. The climate crisis is already evident in droughts, extreme temperatures, and threats to biodiversity and human health—all exacerbated by industrial agriculture and pollution. A paradox in public opinion further complicates this situation: 72% of Czechs support nature conservation, and while 52% view addressing climate change as urgent, and another 25% believe it must be tackled within this decade, most of them neither understand nor trust the measures of the Green Deal, and political leaders have failed to pursue decarbonization. Instead, they exploit fears of the green transformation and neglect to adequately explain to citizens the climate crisis and necessary green transformation (Krajhanzl et al.).

Today, bark beetle infestation is one of the most visible environmental problems exacerbated by both climate change and past environmental policies. The vast spruce forests that dominated much of the country and contributed to its Central European fairytale charm—alongside picturesque villages, castles, and churches—have withered under its onslaught. While scientists report positive developments, such as biodiversity burgeoning in some regenerating areas, the public feels the ongoing loss of forests.

Psychologist and environmentalist Barbara Klocová's documentary *Lesní žal* (*The Grief of the Forest*, 2023, 26 minutes) draws on thirty-one personal interviews with children and adolescents from the Bohemian-Moravian Highlands. These accounts reveal how young people affected by this ecological crisis view it. The documentary combines intimate, personal narratives with a broader analytical perspective as part of a larger project examining the environmental crisis's emotional and psychological effects on local communities. It provides a dual experience to audiences as both an emotional exploration of loss and resilience and a critical reflection on the broader social and psychological dimensions of ecological crises.

The damage wrought by the bark beetle has transformed places familiar to many young people beyond recognition. Beloved spots where they once played, daydreamed, and relaxed vanished. Although the term *solastalgia*—"the feeling of being homesick while you're still at home" (Albrecht 45)—is not stated in the documentary, the emotions expressed by the young interviewees reflect its essence: a sense of distress caused by witnessing one's home environment transformed by environmental crises.

This documentary does not portray forests as wilderness areas or ecosystems separate from humans; instead, it focuses on why young people value them and mourn their loss. When asked about their experience in the forest, the interviewees describe feelings of peace, freedom, and happiness, as well as the joy of breathing clean air and seeking refuge in the "green shadow." Seeing these places devastated and scarred by machinery evokes strong emotional responses, often expressed through metaphors: participants compare the affected areas to the surroundings of Chornobyl, opencast mines that scar industrial regions in the Czech Republic, or landscapes subjected to war raids. One girl offers an intensely emotional image, describing a shadow closing the door where a beautiful forest once stood, leaving her to walk alone into a wasteland. Some participants reflect on their altered emotional state, with one noting, "I feel more negative than I used to." Adult interviewees liken the environmental changes to apocalyptic scenarios or the death of a close friend.

However, the documentary does not dwell solely on loss; it also explores possible remedies for both nature and its human inhabitants. The key to adaptation, as presented in the film, lies in actively caring for the landscape. By seeding and protecting new forests in the devastated area, young people contribute to the restoration of the landscape while learning to cope with loss. As their teacher explains, the restoration project offers children hope, showing them that not all changes must be for the worse. Over several years since the start of the project, participants have witnessed the emergence of young wilderness.

An essential element of many ecocritical approaches is what Scott Slovic calls the "polymorphously activist" ("The Third" 7) tendency, which arises from scholars' and educators' urgent desire to contribute to the green transformation of the global community. This tendency is grounded in the belief, shared by many ecocritics, that "change in cultural values can lead to less destructive ways of living" (Clark 4). The adult initiators of the project express hope that their efforts will have a lasting impact on the younger generation. They aim to instill in them the understanding that forests do not have to disappear entirely, to inspire awareness of the power individuals hold to shape the landscape and to foster a sense of responsibility through time spent in nature and the physical effort invested in rebuilding it. The children address the forests at the documentary's end, wishing them good luck and asking them "not to write us off." One boy, however, directs his plea to the people, the documentary's audience, urging viewers to "slow down" and reflect on their actions.

In the documentary, human corporeality and emotional needs for contact with nature are understood as natural and mutually beneficial. Individual actions are not destroying the forest; human love and care for nature offer hope for restoration and protection through physical stewardship and, above all, by creating a counterbalance to a profit-driven system that has facilitated environmental disaster. Physical engagement with forests is foundational for fostering intimacy with the landscape and embracing the urgency of environmental protection in accord with the ethical dimension of transcorporeality.

Ecoadaptation Studies: Ecocriticism, Adaptation, and Environmental Narratives

Both ecocriticism and adaptation studies are interdisciplinary and dynamically evolving fields. While their openness is a strength, it also presents methodological challenges that require precise theoretical framing. These two disciplines can intersect in analyzing the transformation of environmental narratives across media[4]—whether through the diachronic study of adaptations of classical environmental works or by examining how contemporary adaptations of older texts incorporate ecological themes and approaches—thereby constituting a possible subfield of ecoadaptation studies. Robert Geal suggests studying how ecophobic attitudes and their representations are "adapted through time, across cultures, between texts, and so on" (Geal 6). A shared focus on materiality provides a further connection: material ecocriticism[5] and adaptation studies intersect in object-oriented ontology, which Geal identifies as an inspiring framework for both fields (3). Kyle Meikle extends this connection to a practical dimension by exploring the greening of film production and the sustainability of adaptation studies, asking, "What is the journal of *Adaptation*'s carbon footprint?" (267). Such approaches bridge academic research and environmental practice.

In their recent book Intermedial Ecocriticism: The Climate Crisis through Art and Media (2023), Jørgen Bruhn and Niklas Salmose propose a productive synthesis of intermedial studies and ecocriticism. Their approach highlights the value of comparative media analysis, offering what they call "significant precision in pinpointing the exact affordances of the general media type and the specific media product" (49). This framework proves especially useful when analyzing media transfers and the ways ecological discourse is mediated across various forms. Within this context, adaptation studies shares several key concerns with intermedial studies—particularly regarding how different media

[4] Ecocriticism has evolved beyond its initial focus on literature to encompass the exploration of environmental themes across various media. A notable illustration of this shift is Alexa Weik von Mossner's *Affective Ecologies: Empathy, Emotion, and Environmental Narrative* (2017), which offers insightful analyses of film and literature.

[5] In the introduction to the collective monograph *Material Ecocriticism* (2014), editors Serenella Iovino and Serpil Oppermann define the discipline as exploring the connections between material processes and discourses, emphasizing the inseparability of matter and meaning. In this framework, matter is understood as an active participant in the world, possessing its own narrative potential—what they term "storied matter." Material ecocriticism thus investigates how materiality shapes our perceptions of the environment, focusing on how the substances and forces of the natural world interact with human activities, culture, and artistic expression. It conceptualizes the relationships between humans and things as complex networks of mutual interactions. At its core lies an interest in the influence of nonhuman entities—substances, plants, animals, and ecosystems—on environmental narratives (1–2).

forms interact and generate new meanings. Yet the theoretical relationship between the two disciplines remains unresolved. Adaptation studies tends to view adaptation as a cultural principle—fundamentally fluid, hybridized, and inseparable—where distinctions between elements are porous or dissolved. In contrast, intermedial studies resists such total hybridity; it assumes that meaningful cultural analysis depends on the partial differentiation of media forms. Without such distinctions, intermediality would lose its analytical grip. These are two cognitive paradigms: one grounded in movement, the other in structure. They may appear aligned but ultimately remain incompatible (Bubeníček, "Adaptační" 110). Ecoadaptation refers to the process through which ecological themes and environmental concerns are translated, reinterpreted, or transformed across different media forms—especially narrative and visual media. While the term can be understood broadly, this case study adopts a narrower definition: it focuses on the adaptation of environmentally oriented narratives that foreground ecological issues, with particular attention to how ecological discourse is reshaped within the medium of documentary film. In this sense, ecoadaptation also includes a reflection on how ecological narratives circulate across media and how they are received in participatory artistic practices. Returning to Christof Mauch's observations, ecoadaptation invites a diachronic perspective that acknowledges the evolving historical relationships between humans and the environment. It thus reflects not only a transformation in form and medium, but also a reconfiguration of how ecological agency is imagined across time.

Ecocriticism often draws attention to overlooked aspects of cultural texts, such as reduction of nature to a mere backdrop, while critiquing implicit assumptions about the relationship between nature and culture. Adaptation studies can deepen this critique by investigating how environmental narratives are reshaped across media and how ecological themes are made visible (or invisible) in the new medium. Together, these fields offer a framework for critically engaging with environmental issues through cultural representations. Helena Hughes defines *eco-documentary* as a subgenre of documentary filmmaking focused on ecological themes, exploring the relationships between humans, nonhuman species and the environment (Huges 8–9). Eco-documentaries adapt the broader ecological discourse, the history of ecological movements and other related topics. Documentaries engaging with the ecological discourse draw from various sources, such as environmental philosophy, political debates, activism, and the changing scientific knowledge in ecology. Documentaries addressing the environmental crisis thus can serve as tools for educating and informing the public on environmental issues.

This study illustrated how environmental discourse in Czech society has evolved from the repressive context of state socialism to contemporary neoliberal capitalism. The selected documentaries emphasize individuals' emotional and cultural ties with their natural environments. Portraying nature as an active participant rather than a passive backdrop challenges traditional environmental narratives, revealing how media products reflect, distort, or amplify ecological meaning. Viewed through the lens of ecoadaptation studies, these documentaries illustrate how environmental narratives are not only mediated but transformed across historical, cultural, and media contexts. By drawing on diverse aesthetic strategies and ecological concerns, they offer rich case studies for exploring how adaptation can serve both as a cultural process and as an environmentally engaged practice. All three documentaries examine the nature-culture relationship shaped by the ecological crisis, using distinct methods to engage viewers—poetic expression, mockumentary, and direct appeal—to underscore the urgency of reevaluating individual and collective priorities. Hapl's documentary portrays forests as aesthetic and cultural treasures, as well as indispensable pillars of ecological stability, without which the landscape and humanity itself cannot survive. *Oil Gobblers* subverts any perceptions of prosperity tied to coal mining by exposing immense landscape devastation. In contrast, *The Grief of the Forest* emphasizes the positive relationship between humans and nature, with a local focus on care for specific places tied to a broader understanding of the global causes of the environmental crisis and the need for collective action at both community and international levels. From the perspective of transcorporeality, which challenges the habitual separation of body and landscape, the three documentaries offer different positions: In *Forests Die Silently*, the mere presence of human bodies and needs poses a problem for the survival of the forests. In *Oil Gobblers*, the connection between the well-being of the landscape and the bodies is deliberately disregarded. In contrast, in *The Grief of the Forest*, hope for the flourishing of the forests comes from a human physical, emotional, and ethical connection to the forest. In this way, the selected documentaries exemplify how ecoadaptation can respond to environmental discourses—adapting not only stories but also modes of aesthetic frameworks. They invite us to consider adaptation as an ecological process, capable of forging new relationships between media, meaning, and the material world.

Works Cited

Alaimo, Stacy. "Trans-Corporeality." *Posthuman Glossary*, edited by Rosi Braidotti and Maria Hlavajova, Bloomsbury Academic, 2018, pp. 435–8.

Alaimo, Stacy. "States of Suspension: Trans-corporeality at Sea." *Interdisciplinary Studies in Literature and Environment*, vol. 19, no. 3, 2012, pp. 476–493.

Alaimo, Stacy. *Bodily Natures: Science, Environment, and the Material Self*. Indiana UP, 2010.

Albrecht, Glenn. "'Solastalgia': A New Concept in Health and Identity." *PAN: Philosophy Activism Nature*, no. 3, 2005, pp. 44–59.

Bruhn, Jørgen, and Niklas Salmose. *Intermedial Ecocriticism: The Climate Crisis through Art and Media*. Lexington Books, 2023.

Bubeníček, Petr. "Adaptační studia v recyklační fázi kultury." *Protřepat, nemíchat! Mezi literární vědou a kulturálními studii*, edited by David Skalický and Jan Wiendl, Univerzita Karlova, Filozofická fakulta, 2022, pp. 104–114.

Bubeníček, Petr. "Irony and Catastrophe: The Environmental Crisis in Jan Svěrák's Oil Gobblers." *Ekphrasis: Images, Cinema, Theory, Media*, vol. 33, no. 1, 2025, pp. 13–26. Ecological Emergencies Across Media, edited by Heidi Hart and Jørgen Bruhn.

Clark, Timothy. *The Value of Ecocriticism*. Cambridge UP, 2019.

Geal, Robert. "Towards an Ecocritical Adaptation Studies." *Adaptation*, vol. 16, no. 1, 2023, pp. 1–12.

Glotfelty, Cheryll. "Introduction: Literary Studies in an Age of Environmental Crisis." T*he Ecocriticism Reader: Landmarks in Literary Ecology*, edited by Cheryll Glotfelty and Harold Fromm, U of Georgia P, 1996, pp. xvii–xxxvii.

Hughes, Helen. *Green Documentary: Environmental Documentary in the Twenty-First Century*. Intellect, 2014.

Iovino, Serenella, and Serpil Oppermann. *Material Ecocriticism: Materiality, Agency, and Models of Narrativity*. Indiana UP, 2014.

Jehlička, Petr, and Joe Smith. "Trampové, přírodovědci a brontosauři: předlistopadová zkušenost českého environmentálního hnutí jako předzvěst ekologické modernizace." *Soudobé dějiny*, vol. 24, no. 1–2, 2017, pp. 78–101.

Kahana, Jonathan. "Editor's General Introduction." *The Documentary Film Reader: History, Theory, Criticism*, edited by Jonathan Kahana, Oxford UP, 2016, pp. 1–11.

Krajhanzl, Jan, et al. "České klima 2024." *Institut2050*. institut2050.cz/wpcontent/uploads/2024/10/ceskeklima2024.pdf. Accessed 14 January 2025.

Lesní žal (*The Grief of the Forest*). Directed by Barbara Klocová, Czech Republic 2023.

Lesy umírají tiše (*Forests Die Silently*). Directed by Václav Hapl, Krátký film Praha, Czechoslovakia 1978.

Mauch, Christof. "The Ecocene: A History of Ecological Ideas." *Rachel Carson Center Perspectives*, no. 2, 2010, pp. 1–16.

Meikle, Kyle. "Is Adaptation Studies Sustainable?" *Adaptation* vol. 14, no. 2, 2021, pp. 260–273.

Nichols, Bill. *Introduction to Documentary*. 2nd ed., Indiana UP, 2010.

Ropáci (*Oil Gobblers*). Directed by Jan Svěrák, FAMU, Czechoslovakia 1988.

Slovic, Scott. "The Third Wave of Ecocriticism: North American Reflections on the Current Phase of the Discipline." *Ecozon@: European Journal of Literature, Culture, and Environment*, vol. 1, no. 1, 2009, pp. 4–10.

Slovic, Scott. "Ecocriticism 101: A Basic Introduction to Ecocriticism and Environmental Literature." *Pertanika J. Soc. Sci. & Hum*, vol. 23, 2015, pp. 1–14.

Vaněk, Miroslav. *Nedalo se tady dýchat: Ekologie v českých zemích v letech 1968–1989*. Praha Maxdorf, 1996.

4

The Gloom in the Attic: Fungal Networks as Ecofeminist Adaptation Studies

Cat Champney and Juliana Jones-Beaton

Below every forest, there is an invisible world composed of fungi. Mushrooms provide the tiniest evidence of something greater underground: a living mycelial network. This network, made up of hyphae, tiny tube-like threads, comprises a complex system of nutrition, information, and communication that connects the forest in a web of relationships. More than ninety percent of plants depend on mycorrhizal fungi to survive, and as Merlin Sheldrake writes in *Entangled Life,* fungal networks are living things, not inert resources (4) and a mycelial network is a "map of a fungus's recent history" (53). And, though fungi in popular media are often framed as parasitic, siphoning the host's life form for their own benefit, fungal symbiosis is most often mutualistic. When a plant and a fungus enter a symbiotic relationship, the mycelial network becomes a

C. Champney · J. Jones-Beaton (✉)
University of Delaware, Newark, DE, USA
e-mail: jtrue@udel.edu

C. Champney
e-mail: champney@udel.edu

P. Demory (ed.), *Ecoadaptation*, Palgrave Studies in Adaptation and Visual Culture, https://doi.org/10.1007/978-3-032-09847-4_4

mycorrhizal network: mycorrhiza meaning "fungus" and "root." Mutualistic symbiosis, by definition, benefits both fungus and plant: a mycorrhizal network increases resistance to disease and adverse situations and bolsters nutrient and water intake. Mycelial networks in forests have been referred to as the "wood wide web," as the threads connect trees and plants in an internet-like phenomenon, a symbol of forest collaboration.

Adaptations, like fungi, have been framed as contagious, life-sucking parasites. Such moralistic language describing adaptations has persisted, despite scholars like Robert Stam[1] calling for new tropes, adaptations remaining under cultural fire for their treatment of an underlying "pure," "original" text. However, like fungi, adaptations participate in networks that preserve the lives of source texts and ensure they survive (and potentially thrive) in new environments. Adaptation networks promote a symbiotic relationship between an adaptation and its many sources; adaptations, as Lissette Lopez Szwydky has said, "keep texts alive" (25). Silvia Moreno-Garcia's postcolonial Gothic novel, *Mexican Gothic* (2020), capitalizes on the similarities between adaptation and fungi in popular culture and, in the process, models an ecofeminist approach to adaptation studies. Ecofeminism converges feminist and ecological concerns, generally seeking new ways to understand relationships between gender and the environment. While adaptation studies have utilized gender and environmental theories separately, there has been little consideration of what, exactly, an ecofeminist approach to adaptation might look like. Moreno-Garcia's novel is not the first postcolonial Gothic to utilize adaptation—Jean Rhys' *Wide Sargasso Sea,* perhaps, models Moreno-Garcia's general approach—but her use of fungal networks to illustrate textual adaptation demonstrates new ways in which contemporary adaptors embed adaptation networks in their texts' environments, making adaptation a part of the mycorrhizal network and, by proxy, a part of the textual "wood wide web."

Specifically, we define the relationship between *Mexican Gothic* and "The Yellow Wallpaper," one of the texts Moreno-Garcia adapts, as a holobiont. In biology, holobionts—"holo" for "whole" and "biont"

[1] See, for example, Robert Stam's introduction to *Literature and Film: A Guide to the Theory and Practice of Film Adaptation*, 2004, edited by Robert Stam and Alessandra Raengo.

meaning a unit of living matter—are the sum of the parts, referring to the ecosystem or assemblage composed of multiple organisms. Lichen, one of the most familiar holobionts, is composed of two distinct organisms: algae (or cyanobacteria) and fungi. The research on how these individual organisms interact is vast, with most scientists agreeing that lichen symbiosis is mutualistic: the fungi provide protection and a physical place to grow, while algae photosynthesize food for the fungus. As both organisms give to one another, they similarly take: lichens take up physical space and fungi take the nutrition they need to survive. Holobionts are in no way symbols of peace: for lichens, one organism—particularly the fungi—may benefit more from the relationship than the other, and many holobionts contain parasitic relationships as well. Regardless, the organisms in holobionts have evolved to interact and to make something new. To think of *Mexican Gothic* as the fungal organism and "The Yellow Wallpaper" as the algal organism means to envision the ways in which each text-as-organism—and the networks between them—form a holobiont and, thus, use each other. In lichens it is difficult to tell which part is algae and which is fungi, because they are inseparable pieces of a whole. Similarly, *Mexican Gothic* and "The Yellow Wallpaper" give and take from the other, creating a complex fungal adaptation network.

Moreno-Garcia's novel has far more than one source—praise for the novel references Daphne du Maurier's *Rebecca,* Charlotte Brontë's *Jane Eyre,* Edgar Allan Poe's "The Fall of the House of Usher," and H.P. Lovecraft's canon at large, among other popular Gothic stories. However, this chapter isolates *Mexican Gothic's* relationship with Charlotte Perkins Gilman's "The Yellow Wallpaper" to highlight the text's fungal relationship to the "women's" literary canon and (eco)feminist scholarship. Moreno-Garcia utilizes contemporary knowledge of the environment and, specifically, mycology, to highlight how Gilman's fungal wallpaper has spread across texts and time, adapting to new environments. *Mexican Gothic* follows Noemí Taboada, a young Mexican socialite, as she visits her cousin, Catalina, who recently married into the wealthy, old-fashioned Doyle family. Virgil Doyle, her new husband, is the heir to Howard Doyle, the patriarch of High Place and a proponent of eugenics, exemplified as he openly obsesses over "fit and unfit people" and "superior and inferior types." Though set in Mexico in the 1950s,

High Place mimics a nineteenth-century Gothic novel: the house is "like something out of a Gothic tale" (23). As an anthropologist, Noemí is quickly intrigued by the house and by its environment, specifically the wallpaper: "The wallpaper was not green," Noemí observes, "Not even close to green; it was a muted pink, the color of faded roses, with ugly *yellow* medallions running across it.... This was hideous, and when she closed her eyes, the yellow circles danced behind her eyelids, flickers of color against black" (47). Like the protagonist in "The Yellow Wallpaper," Noemí realizes that information is spread through the walls, and fungus (the "gloom") is instrumental in her survival. Infectious in nature, the yellow circles—traveling from the paper to Noemí's body—allegorize the spread of Gilman's narrative, "The Yellow Wallpaper," to *Mexican Gothic.*

In "The Yellow Wallpaper," Jennie, the protagonist's sister-in-law, finds "yellow smooches" on the protagonist's and her husband's clothing (8). These yellow smooches spread from the walls to the narrator/protagonist's mind. These "yellow smooches" also spread from the physical paper of Gilman's narrative to the minds of critics and readers, to the mind of Moreno-Garcia, and, finally, to the material of *Mexican Gothic*. After all, "one way back to 'The Yellow Wallpaper' is through the yellow wallpaper itself" (Lanser 424). As such, the yellow circles behind Noemí's eyes do not stem just from the walls of High Place but also the walls of Gilman's narrative, traveling through the fungal adaptation network. These "yellow smooches" are not deadly, as they form a holobiont with other texts' fungi, such as the fungus growing in *Mexican Gothic.* Through the holobiont, the two texts do not exist in perfect harmony, reflecting the tumultuous, though necessary, relationship between source texts and adaptations.

Noemí is not the first literary character affected by yellow smooches. The wallpaper that drives the female protagonist mad in "The Yellow Wallpaper" contains a fungal pattern: "The outside pattern is a florid arabesque, reminding one of a fungus. If you can imagine a toadstool in joints, an interminable string of toadstools, budding and sprouting in endless convolutions, why, that is something like it" (Gilman 41–42). Gilman's narrator becomes increasingly convinced that the patterns of the wallpaper in her room are hiding women within. She is insistent that "it is like a woman stooping down and creeping about behind that pattern" (Gilman 39). In mycophobic[2] popular culture, fungi often appear as an entity to fear: a contagion, a parasite, or a monstrous villain that strips human characters of their agency and autonomy, if not their lives—in a way, this form of popular culture treats fungi like adaptations. These are often metaphorical roles, as the fungi symbolize ideological "spread" relating to class, gender, and race. By "toadstool in joints," Gilman refers to uncontrollable poisonous mushrooms, the consumption of which can cause liver failure and, by proxy, turn human skin yellow. Her text, then, assumes a mycophobic attitude toward fungi even as it survives due to the fungal networks it shares with other texts, such as *Mexican Gothic*.

The fear of "yellow" skin in Gilman's story implies not only jaundice as a side effect, but also embedded nineteenth-century racial ideologies. These ideologies have been, for the most part, masked by the short story's prestige in the literary canon. "The Yellow Wallpaper" has been extensively theorized as a feminist text, occupying a central role in Sandra Gilbert and Susan Gubar's seminal *Madwoman in the Attic* (1979) and, as a result, a foundational "place of privilege" in the second-wave feminists' female literary canon. Gilman's protagonist and her "reading" of the yellow wallpaper are powerful metaphors for women's (and feminists') literacy. Critics like Susan Lanser, though, have responded to the overwhelming emphasis on feminism in Gilman's text by pointing out subjects that have been overlooked, such as the color of the wallpaper, and the ways in which these subjects endorse patriarchal ideologies and contradict third-wave feminist values, primarily intersectionality and the significance of race to feminist theories. Lanser argues that the wallpaper in Gilman's tale cannot be separated from the cultural and political context of the "Yellow Peril," a form of Anti-East Asian racism that dominated America during Gilman's life and time as a writer. "In such a cultural moment," Lanser writes, "'yellow' readily connoted inferiority, strangeness, cowardice, ugliness, and backwardness.... Associations

[2] Mycophobia: the irrational fear of mushrooms or the belief that any fungus and its environment are toxic (Prime and Broussard).

of 'yellow' with disease, cowardice, and worthlessness, uncleanliness, and decay may also have become implicit associations of race and class" (238).

Gilman's yellow wallpaper is suggestive of racial anxieties, but, as Lanser has highlighted, also of toxic reproduction. Gilman mobilizes fungi as "yellow" contagions, or "smooches," that "stain" everything they touch. The yellow stains reflect real fears about contaminated wallpaper in the Victorian era. As Agnes Malinowska asserts, "enthusiasts of the new home economics movement were urging women to take down their wallpaper, which they considered to be a breeding ground for deadly microbes–like fungus–that had recently been described by the germ theory of disease" (271). Malinowska extends her reading of the "breeding ground" to claim that Gilman's text displays a "world of women trapped by uncontrollable—'interminable'—toxic reproduction" (271). This "toxic reproduction" is not limited to characters and objects in Gilman's story, though, as the "yellow smooches" have spread outside the text. The fungal wallpaper has become a popular motif in stories about women's struggles in patriarchal systems, particularly those that portray postpartum depression or madness in the wake of marriage and sexual violence. And the motif of yellow, or fungal, wallpaper is an indicator of adaptations of Gilman's text, even those that may take a subtler approach to retelling her story. Julie Grossman has argued, for instance, that Todd Haynes' *Safe* (1995) is a "quiet" adaptation of "The Yellow Wallpaper," and she cites Carol White's recollection of the yellow wallpaper from her childhood bedroom early in her argument, framing the appearance of the paper as a foundation for the texts' more complex engagement (106). Therefore the "yellow smooches" that infect Gilman's protagonist and real Victorian women continue to infect adaptations. However, rather than repeat Gilman's mycophobic lens, Moreno-Garcia embraces these yellow smooches and renegotiates the relationship between the women and the wallpaper, enabling Noemí, her cousin, and generations of Doyle women to overcome the Doyle patriarch's abuse through a fungal network.

Mexican Gothic is infested with mushrooms. Fungi permeate the pages of the novel: Noemí eats mushrooms for dinner, Francis gathers mushrooms in the cemetery, the servants wash them, Francis draws them,

Noemí painstakingly records the mildew she finds in her room. Mushrooms are everywhere, but the most important fungus is not visible, but woven into the walls and underneath High Place. This fungus is, unlike Francis' spore prints, entirely alive. High Place, and the characters inside, rely fully on the construction and architecture of the fungi, called "the gloom." The gloom acts as a conduit for information and, when appropriated, fuels immortality to the patriarch of the Doyle family:

> The fungus, it runs under the house, all the way to the cemetery and back. It's in the walls. Like a giant spider's web. In that web we can preserve memories, thoughts, caught like the flies that wander into a real web. We call that repository of our thoughts, of our memories, *the gloom*. (211)

Presented in this way, the gloom is largely portrayed as negative, something feared and incomprehensible. However, the true villain in High Place is not the fungal network, but the Doyles' fear of racial "contagion" and contamination. The Doyles consider the mycelial network to be an inert thing, a pliable resource, not a living process. As such, they are ignorant of the possibilities of the fungus past what it can grant them in the form of immortality. The fungus that constructs the gloom, illustrated through memories and fragments, was originally growing inside a cave in Mexico, used by native people to heal their wounds and "preserve their health." Howard Doyle exploits the fungus to do "so much more": to achieve eternal life (206). Realizing that he needed Indigenous blood, he impregnated one of the women to later eat the child. This act of violence is one stage in Doyle's colonialist agenda: he effectively consumes an entire community to gain sole access to the resources he seeks to exploit: the fungus. Doyle transplants the fungus to High Place, ripping the fungi from its home, violently and without regard, as an act of colonial violence in and of itself. The fungi reside next to many plants that Doyle has taken from England, embedding them in soil from his homeland, and this garden of sorts perfectly exemplifies the Doyles' ecological violence and misunderstanding of the world: they believe they can control a fungus through English soil, effectively "tricking" the fungi to behave in certain ways, regardless of locale.

Like the fungi, the women in High Place are treated as disposable resources. They are bodies to be transplanted and used to sustain the fungus and provide children to the genealogical line. Howard Doyle often emphasizes the fertility of women as their only use and he tells Noemí that Agnes, his first wife, "was fruitful," as "a woman's function is to preserve the family line" (75). During transmigration, the act of Howard's consciousness slipping into a new shell of a human being, he refuses to migrate into a woman's body, instead targeting his male heir. A woman's body is a thing to be exploited, but only in certain, set, ways, determined by the immortal patriarch of the house. This treatment of the women and the fungi in High Place as individual, distinct tools results in an ignorance of the network of women that experience the violence of High Place. In treating each of the women in the same demeaning way (Catalina, Noemí, Agnes, Ruth, Alice) the Doyles unite them, threading together their experiences. Like the women in High Place, the mycelium that connects them is similarly not inert or static. Sheldrake writes that mycelium "describes the most common of fungal habits, better thought of not as a thing but as a process: an exploratory, irregular tendency" (6). Mycelium does not just refer to the physical threads of hyphae, but rather the living and seeking process of fungal growing. The Doyles' critical mistake is that they do not consider the agency of the women inside High Place, nor do they consider the agency of the mycelium that sustains it.

When Noemí arrives at High Place at the start of the novel, she enters the adaptation network directly, as the Doyle ancestral home adapts the setting of "The Yellow Wallpaper." Like the protagonist in Gilman's narrative, Noemí must "read" the fungal paper to free herself and the women behind the walls. Her engagement with the gloom solidifies this network as she begins to realize what, exactly, she is reading. Like the wallpaper in "The Yellow Wallpaper," the gloom causes Noemí to mimic her cousin Catalina's symptoms: she experiences a lack of appetite and bad dreams.[3] Noemí's position as a reader of "The Yellow Wallpaper" becomes more transparent as she views the wallpaper in the same

[3] Gilbert and Gubar classify these symptoms as the result of traditional training for matrimony in the nineteenth century—a low appetite and weakness were considered ideal "feminine" traits (54).

way she views texts, like drawings: "she could now see the wallpaper, its drawing of flowers, faint traces of mold marring its surface" (119). Noemí's reading highlights how a motif, such as the yellow wallpaper, creates an adaptation network and, like a fungus, participates in a cross-textual mycelium network, simultaneous and symbiotic to its adaptation network.

Though Noemí believes Catalina to be mentally ill at the start of the narrative, Catalina reveals the fungal (and, thus, adaptation) network to Noemí: "There're people in the walls. There're people and there're voices. I see them sometimes, the people in the walls. They're dead" (85). Catalina's own reading preference for fairy tales and her supposed "happy ending" with Virgil positions her as a more naïve reader, a reader who, like many early scholars of "The Yellow Wallpaper," overidentify with a text and its protagonist, even if the text does not speak to one's experience. The contrast between Noemí and Catalina's reading preferences aligns with third-wave feminist criticisms of "The Yellow Wallpaper" and the academy. Moreno-Garcia adapts this line of thinking and transforms Noemí into a critic, whose reading of the wallpaper in High Place enables her to escape from confinement rather than fall victim to it. Confinement, here, relates to her entrapment both in High Place and in a literary canon that does not speak to her experiences as a Mexican woman.

Catalina and Noemí are also two distinct takes on the "classic" madwomen in Gothic literature. The figure of the madwoman has been broadly understood as a double for the woman writer, though Moreno-Garcia repositions Noemí as an anthropologist, studying humans in relation to their environment. Gilbert and Gubar theorize that women writers "often envision an 'outbreak' that transforms their characters into huge and powerful monsters," and create tension in writing through exploration of "the psychic split between the lady who submits to male dicta and the lunatic who rebels" (116). To Gilbert and Gubar, such lunacy is catching, and Gilman's "The Yellow Wallpaper" is "*the* story that all literary women would tell if they could speak their 'speechless woe'" (119). They envision Gilman's protagonist as the reader and writer of the wallpaper, like women who enter High Place, trapped by patriarchal stories and freed by her own revisions to their bar-like patterns. The "creeping" woman's escape from "the numb world behind the patterned walls of the text," Gilbert and Gubar argue, "was a flight from [infection] into health" (122). *Mexican Gothic* challenges this, as the gloom is not just an infection to survive and flee from, but a vital source of information for the surviving women at the end of the story. This is what transitions Moreno-Garcia's feminism from the second-wave themes in Gilman's story to third- and fourth-wave ecofeminist values.

When Noemí wonders why the gloom has been showing her faint, jumbled echoes of women, Francis tells her that "it wouldn't be intentional. I guess it's in its nature" (211). Sheldrake describes fungal hyphae that possess directional memory—when placed in a maze, mycelial threads learn and remember which pathways are blocked off via sensory input—and become a wealth of information and memories, much like the gloom in High Place (47). Francis dismisses fungal intent or agency; however, Noemí can see through the gloom in moments of intense danger, through her connection to women who seek to protect her. When Noemí encounters Ruth, Howard's niece who died after attempting to kill her family and destroy the evil of High Place, in the gloom, she is not quite a ghost or a memory, but rather the "mind of Ruth" (233). In addition to Ruth, Noemí recognizes that the gloom also held "other minds, bits of persons, hidden underneath the wallpaper, but none as solid, as tangible as Ruth" (236). Indeed, the information that allows Noemí to escape High Place comes from the gloom and, by proxy, the mycelial (and adaptation) network. Likely without it, Noemí would have expired in High Place like many others before her. The gloom presents her with the history of High Place through the memories of shattered women: primarily Ruth and Agnes, and these women not only give Noemí glimpses of what has happened in the house—for example, Ruth appearing with a shotgun—but constantly warn her, both metaphorically and literally: "open your eyes," a phrase that occurs seventeen times within the text.

Ironically, the Doyles attempt to use fungi in order to preserve their family line: to make the purest and most self-contained genetics, forgetting that fungi, like adaptations, work and *survive* by collaborating and communicating with other organisms. As the contrast between infection and survival suggests, again like adaptations, purity is *not possible* with fungi: the idea of a self-contained adaptation or fungus is indeed antithetical to either one. The women scattered through time in High Place can only interact with Noemí through the existence of the gloom, a side effect of the tight control the Doyles *think* they have over the fungus sustaining them. Francis explains the symbiotic relationship between the fungus and the Doyle family by telling Noemí that "fungi can enter into symbiotic relationships with host plants, Mycorrhiza. Well, it turns out that it can also have a symbiotic relationship with humans. The mycorrhiza in this house creates the gloom" (211). Though Francis describes this as symbiosis, the gloom does not quite fit with the definition of a symbiotic organism. Lynn Margulis writes that symbiosis "is the living together of very different kinds of organisms.... In certain cases cohabitation, long-term living, results in symbiogenesis: the appearance of *new bodies, new organs, new species*. In short, I believe that most evolutionary novelty arose, and still arises, directly from symbiosis" (43, our emphasis). The Doyles attempt to use symbiosis in *Mexican Gothic* not to produce something *new*, but rather to uphold a sort of perverse nostalgia, a view of racially motivated antiquity. The Doyles desire to keep everything *the same,* stagnant and static forever. The family symbol of the ouroboros is, in many ways, a reflection of the Doyles' attempt not only to live forever, but to live forever *a certain way*: repeating an inherently Western, idealized, colonialist, and "pure" version of history. One could argue that the Doyles use the fungus to symbiotically make a new and evolved body, but the Doyles' beliefs and desire for the world to function is the opposite of novelty.

The Doyles think they can use the gloom to keep their bloodline pure and separate, but they forget the most important part of how a fungus works: the webs and networks of mycelium work to *connect*, instead of delimit. When Francis tells Noemí offhand that the memories she absorbs through the gloom are "in its nature," he is right in more ways than he understands, because in both scientific and humanities research, fungi and lichens help to push back *against* the concept of "purity," or as Alexis Shotwell defines, "purity politics" or "purism" (7). Fungi inherently trouble the idea of purity, as their fruiting bodies are constantly contaminating and being contaminated, or perhaps more accurately, *collaborating*. At the root of purity politics is what Alexis Shotwell has called defensive individualism (11) and what Scott Gilbert refutes in the oft-cited title of his paper: "we have never been individuals." Defensive individualism distinguishes humans as self-sustaining and independent beings—a distinction which is not only false but, as Shotwell and others argue, cuts off the possibilities of collective action or recognizing oneself as only one part of an ecosystem or assemblage. This is applicable not just to fungi, but to adaptations too. Adaptations do not exist in solitude; they are by nature *impure*. They are constantly referencing, complicating, rewriting, and responding to other texts, emphasizing the positive aspects of contamination and collaboration through writing. So, too, do adaptations *adapt* and evolve, returning to ideas and source texts to consider new perspectives and ways of thinking.

Writers like Charlotte Perkins Gilman have traditionally weaponized fungi to "spread" ideological fears, while feminist adaptors have recuperated fungi as a tool to empower marginalized people and texts, as they destroy and reconstruct canons that seek to exclude them. This is particularly true in *Mexican Gothic*, as Moreno-Garcia challenges parasitic representations of fungi in canonical short fiction by the likes of Gilman and demonstrates how mycorrhizal networks establish a symbiotic relationship between women across texts, and texts across canons—even when those texts may contain harmful ideologies, as is the case with Gilman's story. Mycorrhizal networks can be a metaphor for adaptation networks, as they "infect" exclusive canons to produce more diverse stories, enabling all adaptation(s) to survive. Though mycophobic texts frame fungus as harmful, fungi actually destroy canon boundaries and force texts like "The Yellow Wallpaper" to embrace their critiques, destabilizing the notion of an unwavering feminist text in favor of diverse interpretations and adaptive textual categories and identities. High Place is crumbling, not only from age, but from the impact of the gloom and the criticisms against canonical stories like Gilman's. By incorporating Gilman's racial anxieties into Noemí's reading of the text, Moreno-Garcia highlights the difference between a "universal" women's experience and a racialized one. *Mexican Gothic* might be considered an "impure" adaptation, infected by a series of other texts such as Gilman's, carrying symptoms like Gilman's yellow paper and its racial ideologies. However, as *Mexican Gothic* demonstrates, adaptations can use fungal networks to promote change and symbiosis, thus their "impurity" is a strength.

In his call for ecocritical adaptation studies, Robert Geal suggests that, before further development, scholars should consider "how adapted texts communicate negative attitudes towards the environment, as well as how they communicate positive attitudes" (5). Furthermore, "adaptations," he argues, offer "privileged examples of how different moments/cultures relate to these ecological issues, because we can identify the precise moments and characteristics of either consistency or change during texts'... evolutions. We can think of adaptations," he concludes, as laboratories "for the analysis of our species' evolving history of ecophobia" (10). Moreno-Garcia's ecofeminist adaptation models a symbiotic relationship between adaptation and fungi, where each network informs the other. Additionally, reading adaptation(s) as fungal promotes a more nuanced understanding of the relationship between adaptation(s), humans, and the environment, particularly when it comes to anxieties about infections, poisoning, or other alleged environmental "impurities." And, as an ecofeminist postcolonial text, Moreno-Garcia's adaptation "provides a language for re-engaging with the political realities of the post/neo-colonial present, speaking truths that are structurally repressed elsewhere," including institutions like the academy that house adaptation studies (Ilot 22). *Mexican Gothic* speaks environmental and feminist truths, broadening the concept of intersectionality to include the environment.

Mexican Gothic is ultimately hopeful. Unlike Gilman's protagonist, Noemí does not need to "creep" through life as a monstrous figure. Instead, she transcends the text in confidence that she *and* Francis can adapt, even though he may still carry remnants of the fungus from High Place: "The future... could not be predicted and the shape of things could not be divined. To think otherwise was absurd. But they were young that morning, and they could cling to hope. Hope that the world would be remade, kinder and sweeter" (301). Adaptation studies can also be remade, kinder and sweeter, and more representative of intersections between gender and the environment.

Works Cited

Geal, Robert. "Towards an Ecocritical Adaptation Studies." *Adaptation,* vol. 16, no. 1, pp. 1–12.

Gilbert, Scott F, et al. "A Symbiotic View of Life: We Have Never Been Individuals." *The Quarterly Review of Biology,* vol. 87, no. 4, 2012, pp. 325–41.

Gilbert, Sandra M., and Susan Gubar. *The Madwoman in the Attic: The Woman Writer and the Nineteenth-Century Literary Imagination.* 1979. Veritas, Yale UP, 2020.

Gilman, Charlotte Perkins. "The Yellow Wallpaper." 1892. Edited and with an introduction by Thomas L. Erskine and Connie L. Richards, Rutgers UP, 1993, pp. 29–52.

Grossman, Julie. *Literature, Film, and Their Hideous Progeny: Adaptation and ElasTEXTity.* Palgrave Macmillan, 2015.

Ilott, Sarah. "Postcolonial Gothic." *Twenty-First-Century Gothic,* edited by Maisha Wester and Xavier Aldana Reyes. Edinburgh UP, 2019.

Lanser, Susan. "Feminist Criticism, 'The Yellow Wallpaper,' and the Politics of Color in America." "*The Yellow Wallpaper,*" edited by Thomas L. Erskine and Connie L. Richards, Rutgers UP, 1993, pp. 225–256.

Malinowska, Agnes. "Charlotte Perkins Gilman's Fungal Female Animal: Evolution, Efficiency, and the Reproductive Body." *Modernism / Modernity,* vol. 26, no. 2, John Hopkins UP, 2019, pp. 267–288.

Margulis, Lynn. *Symbiotic Planet.* Orion Books, 1998.

Moreno-Garcia, Silvia. *Mexican Gothic.* Del Ray, 2020.

Perkins Gilman, Charlotte. "The Yellow Wallpaper." 1892, *New England Magazine,* edited and with an introduction by Thomas L. Erskine and Connie L. Richards, Rutgers University Press, 1993, pp. 29–52.

Prime, Nate, and Will Broussard. "Mycophobia: The Fear of Mushrooms Reinterpreted." *North Spore,* May 30, 2024. northspore.com/blogs/the-black-trumpet/mycophobia-mushrooms-reinterpreted. Accessed 9 July 2025.

Sheldrake, Merlin. *Entangled Life.* Random House, 2020.

Shotwell, Alexis. *Against Purity: Living Ethically in Compromised Times,* U of Minnesota P, 2016.

Szwydky, Lissette Lopez. *Transmedia Adaptation in the Nineteenth Century.* Ohio State UP, 2020.

Part II

Adaptation in the Age of Climate Change

How do environmentally focused adaptations encourage/discourage or mediate public response to environmental crisis—including climate change? What specific forms/media best represent environmental concerns in the Age of the Anthropocene?

These chapters are connected by their thematic focus on environmental degradation and climate change, but they also examine strategies in form and method for adaptation in the Anthropocene. The first two chapters explore a graphic novel adaptation and a radio play adaptation, examples of media that provide a richer, multi-sensorial experience of environmental crises than traditional textual media, and that may be more environmentally sound as well. The other two chapters explore what happens when oft-told stories are retold in the age of climate crisis. The recycling of ancient texts—such as the Noah story—reveals the way cultural anxieties make their way into new adaptations, but does the focus on environmental concerns in recent adaptations of this story inspire audiences to change or simply ameliorate those anxieties? An adaptation of *Macbeth* filmed on an island succumbing to sea level rise provides stark evidence of climate change—a warning that may or may not be heeded. New media, the recycling of ancient mythologies, the (perhaps accidental) recording of rising sea levels—all are structural adaptations to the current moment.

5

Geo-Echolocating and Bio-Boomeranging: Rayco Pulido Rodríguez's *Nela* as Ecoadaptation of Galdos' *Marianela*

Christopher R. Carter

Spanish realist Benito Pérez Galdós's 1878 novel *Marianela* opens as its ophthalmologist protagonist, Teodoro Golfín, stumbles through the mines surrounding Socrates, the mining village he is struggling to get to. As he wanders—first alone and then guided by the novel's young blind protagonist, Pablo—Galdós devotes significant swaths of Teodoro's trek to realistically visualizing the nightmarish scene of environmental degradation the men traverse. Yet this anthropogenic denudation is only criticized insofar as it hinders the human advancement embodied in the motto Teodoro continuously repeats to himself—"onward, ever onward" (Galdós 74)[1]—in a clear representation of the era's tendency to sever humans from nature in a "historical process usually labeled

[1] All translations in this chapter, unless otherwise noted, are my own. Also, I follow the custom of referring to Benito Peréz Galdós as simply "Galdós"; however, complete bibliographic references for his works can be found under "Peréz Galdós" in the Works Cited.

C. R. Carter (✉)
Universidad de Castilla–La Mancha, Ciudad Real, Spain
e-mail: Christopher.Carter@uclm.es

P. Demory (ed.), *Ecoadaptation*, Palgrave Studies in Adaptation and Visual Culture,
https://doi.org/10.1007/978-3-032-09847-4_5

'progress'" (Heise 167). The novel's perpetuation of this splintering is reiterated when Teodoro first assesses Marianela's appearance after trudging through the mines: "It's a dreadful job, working in the mines. You are covered with mineral-colored dust, you're scrawny and malnourished. This life destroys the most robust of natures" (Galdós 95). Despite his reference to natures in the plural, by elevating the consequences that mining has wrought on *human* nature rather than its degradation of *nonhuman* nature, Galdós' opening passages reveal how humanity's privileged position over nature forms the bedrock on which his novel is built.

But as the consequences of this presumed human superiority become inescapable, calls have been made for a reevaluation of humanity's supposed exceptionalism. Paul Shepard proposes that ecological ways of knowing would contribute to such a recalibration, arguing that "ecological thinking... requires a kind of vision across boundaries" (63) and claiming that "the epidermis of the skin is ecologically like a pond surface or a forest soil, not a shell so much as a delicate interpenetration. It reveals the self enobled [sic] and extended rather than threatened as part of the landscape and the ecosystem" (63). Parallel to this shift toward a more relationally situated understanding of ourselves, Amitav Ghosh argues that confronting the crises of the Anthropocene might force us to "think in images," to make "a departure from our accustomed logocentrism" (83); he suggests the hybridity of graphic narrative to achieve such an end. Indeed, scholars have recently elucidated how the "environmental performance" of graphic narrative creates meaning by relating "images, spaces and texts" (Ghosal and Modak 178) in a way that often "directly engages in ecocriticism" (Dobrin 4); its ability to engage readers and embody the movements and sensations seen on the page (Kukkonen 60–1) might enable a more ecologically situated positioning vis-à-vis the world (Pursall) or foster different animal-human relations (Esben Svelstad 11). Graphic narrative, then, can be an invaluable tool in doing more than just recognizing environmental degradation as a concern in and of itself, something Galdós does not do in *Marianela*; it may give us the boundary-crossing ways of thinking that depict and manifest humans' multifaceted "inness" (Shepard 62) within nature, essential for a radical recalibration of our sense of self.

In this line of thought, this chapter draws on both scenes from the novel and paratextual material that documents the creative processes behind Rayco Pulido Rodríguez's 2013 *Nela: Una Adaptación Gráfica de Marianela de Benito Pérez Galdós* to show how his adaptation prioritizes and experiments with the graphic medium's visuality to create ecologically immersive, "hi-fi" soundscapes and dynamic textual environments that blur the boundaries between human and nonhuman, placing his characters into relational ecologies with the geo- and biospheres that surround them. I also demonstrate how these ecologically relational moves converge as the novel makes a subtle yet clear ecological departure from Galdós' text. This chapter, therefore, concludes that *Nela* serves as a clear example of how graphic narrative's hybridity, and specifically that of a graphic narrative literary adaptation, can indeed reconfigure ecological attitudes and positionings within the original source materials to help recalibrate our sense of self in relation to the environment.

Galdós's *Marianela* centers around a young woman, Nela, who is disfigured as a child, orphaned when her mother commits suicide, deemed unfit to work in the mines as an adolescent, and designated as a guide for Pablo, the congenitally blind man who, unaware of her disfigurement, falls in love with her innocent, naive, and pristine nature. Yet their idyllic courtship begins to unravel when Teodoro arrives: the ophthalmologist encourages Pablo's far more attractive cousin, Florentina, to take Nela under her care in order to "civilize" and "domesticate" her, and he successfully operates on Pablo's cataracts, restoring his sight. Fearing rejection once Pablo beholds her outward appearance, Nela flees, but, upon being forced back to Florentina's home, she witnesses Pablo professing love for his cousin, gushing over her external beauty. When Nela's identity is revealed and she is put on full display for Pablo's gaze, the anguished young woman quickly perishes. Despite the elegant gravestone and elaborate funeral erected for her and funded by Florentina, the narrator informs us that, in the end, the memory of Nela quickly fades.

Unlike Nela's passing into oblivion, however, *Marianela* has enjoyed a notable afterlife as an "eminently adaptable work" (Behiels 767), inspiring three movies, one opera libretto, three theater productions, and two TV series.[2] Pulido, himself a teacher, decided to add his rendition to this list of Galdosian retellings upon seeing how "secondary students were having more and more difficulties with the required reading of *Marianela*" (Ayala 235). This places Pulido's work within the subgenre of graphic and comic literary adaptations whose long history of perceived pedagogical opportunities has often relegated them to a "subservient role—a handmaid's function—in relation to the source material" (Jones 215). Yet, Pulido explains that he did not want to produce "a version for dummies of *Marianela*," affirming that *Nela* is "something different. Each medium has its own language, and the language of comics is among the richest" (Ayala 245). In this rich array of possibilities, the artist has expressed most interest in the medium's visual elements, a preference that clearly lies at the foundation of his creative processes:

> Giving priority to the image and resolving narrative problems with the minimal amount of text possible has always been the most stimulating part of the job. Generating complicity with the reader, trusting their intelligence to construct meaning without overexplaining or being redundant. (Pulido, "Mi panorama")

As these comments indicate, Pulido's commitment to prioritizing the visual elements of graphic narrative over textual ones combines with a desire to engage readers in the text's meaning-creation processes. The application of this principle of visually engaging graphic narratives is central to the scenes in which a subtle yet noticeable ecological undercurrent not present in the original can be detected in *Nela*.

To start with one example, Pulido's process of fully exploiting graphic narrative's visual language is evident in *Nela*'s rendering of Teodoro's initial, harrowing journey through the mines. An entire two-page spread (Figs. 5.1 and 5.2) is dedicated to this narrative moment, which starts with Pablo and his seeing-eye dog, Choto, leading Teodoro through the abandoned quarries (Fig. 5.1 panels 1–3).[3] As they meander through

[2] See Willem's *Adapting Spanish Classics* for more details about each of these respective adaptations.

[3] Unless otherwise noted, all images in this chapter are used with permission of the author, Rayco Pulido Rodríguez, and are taken directly from *Nela*.

the substrate, Teodoro inquires about a song he hears (Fig. 5.1 panel 4 to Fig. 5.2 panel 1) and when Pablo calls out to the unseen singer, his echoes reverberate over the following panels, informing Nela not come meet them (Fig. 5.2 panels 2–5).

Without a doubt, the numerous graphic representations of sound that permeate this scene serve various functions within *Nela*'s narrative, whether by depicting the blind protagonist's dependence on "non-visual senses" (Willem 85) or illustrating how *Nela* "shows" Pablo's differently abled perceptive ability, rather than "tells" about it, as *Marianela* does (Behiels 774). More important for this chapter, however, are how the wildly different visual stylizations in the graphic representation of sound—among the medium's most immediately recognizable elements (Guynes 63)—contribute meaning to the panels and frames. For example, the "tac tac tac" of Pablo's cane (Fig. 5.2 panel 6) forms part of *Nela*'s "comic book soundtrack" (Khordoc), its "special effects" (McCloud 116), a simple "onomatopoetic effect" (Kukkonen 24) whose hollowed-out representation imitates the acoustic characteristics of the sound itself. Or, take the more complex "onomatopeme" (Guynes 63) of the flowery letters of Nela's singing: it is a graphically represented sound that connects "disparate elements of narrative images" and allows readers "to propose and confirm relations between elements in an image" (66), in this case, Teodoro's inquiry about the song and Pablo's subsequent calling out to Nela.

But perhaps the most striking visualization of sound comes in the panels that depict Pablo's shouting and echoes (Fig. 5.2 panels 2–5). While not strictly an onomatopoeic imitation of a natural sound, Pablo's booming echoes epitomize a broader category of graphically represented sound that Dey and Bokil call sound-symbolic words (SSWs): any combination of words, letters, and symbols that not only convey diegetic sound but also indicate details about the time, space, action, emotion, or character identity in a panel or spread (1082). To be sure, the echoes give temporal continuity and verisimilitude to the scene by traversing the row's gutters on the same, slightly oblique angle and depicting a one-panel delay between when the words are uttered in speech bubbles and when they resonate across the background. These echoes also spatially ground the characters atop a craggy pinnacle in the night's darkness,

Fig. 5.1 Teodoro, Pablo, and Choto journey through the mines

Fig. 5.2 The trudging continues

echolocating the protagonists within their geological surroundings. In other words, the SSWs of the echoes emphasize the protagonists' location within the time and space of the mine, which becomes a relational system in which Pablo's human actions are relayed back to the men, dwarfing them in a sight-sound synesthesia that conveys the diegetic soundtrack of the scene but also emphasizes their insignificance within their ecological surroundings.

Interestingly, Pulido's creative process reveals the differing combinations of "onomatopoeia and texture" ("Se busca estilo") he explored to stylize these sounds in a way that replicates what acoustic ecologists have termed pre-industrial, "hi-fi" soundscapes. Such acoustic environments, according to Kendall Wrightson, "offer significant information for the listener, providing cues relating to the physical nature of the environment and expressing its size in relation to the listener" (11); they do so through their heightened ability to echo, absorb, and reverberate the sounds that ricochet around them (11). In his draft version (Fig. 5.3), Pulido inks the echoing panels with black negative space while he frames the men tightly for the entire passage, providing no clues about their location or surroundings; in contrast, the published version replaces this black background (Fig. 5.2 panels 3–5) with cross-hatching to heighten the sensation of the echoes vibrating off the unseen mountainsides, but the shot also widens (Fig. 5.2 panels 3–4), the negative space of the rocky peak now framed by these wavy echoes, the precipice shrunken down to underline the overwhelming size of the soundscape in relation to the men. The published version's visualization of sound, therefore, effectively depicts the scene's hi-fi ecological acoustics and provides the maximum number of details about the scene's ecological relation between the men and the natural geosphere that surround them.

Aside from anchoring his characters in the soundscape of their geosphere, a ten-panel monologue (Figs. 5.4 and 5.5), where Pablo's father, Francisco, contemplates the implications of his son's upcoming operation recalibrates humanity's assumed superiority in the biosphere, blurring the strict distinction presumed to separate the human from the nonhuman. Stylistically, the scene's cartoony, cross-hatched representation (Fig. 5.4 panel 1) is simplified into a high-contrast chiaroscuro silhouette of Francisco (Fig. 5.4 panel 2), an abstraction which is reversed

Fig. 5.3 Draft version of "onomatopoeias and textures." Used by permission of the author, Rayco Pulido Rodríguez, from his personal blog, *Nunca Trabajes Solo*

in the final two panels (Fig. 5.5 panels 4–5). This emptying out, as it were, of the characters' features and representation is a common tactic in graphic narrative, one that allows the artist to universalize, amplify, or make the depicted characters' meaning more malleable (McCloud 30–32). Pulido himself confirms a similar understanding of this tenet in an interview with Linda Willem, stating that "when a character is empty, more abstract, it is necessary to fill it" (82, translated by Willem), a sentiment that clearly aligns with his penchant for producing visually engaging narratives.

Bookended within this back-and-forth abstraction, six realistically stylized panels push Francisco's malleable humanity beyond its bounds, bringing it into direct contact with the nonhuman by combining the words of Francisco's monologue and the image of a realistically represented cow (Fig. 5.3 panels 3–5), thereby developing two parallel tracks of word and image that Scott McCloud defines as a "parallel combination" (54). While *visual* closure is achieved through the zoom-in effect passing across this tier's gutters, no *conceptual* closure is provided, and a bewildering sensation regarding the relationship between cow, fly, and Francisco's comments on inheritance, marriage, and satisfaction emerges. This confusion, however, begins to dissipate when the zoom-in effect pulls back at the beginning of the next page, panning out from an extreme close-up of jagged lines (Fig. 5.5 panel 1) to reveal a work-worn hand (Fig. 5.5 panels 2–3). These depictions of Francisco's hands more clearly intersect with the older man's spoken descriptions of the trials and joys of life that have wrinkled them (Fig. 5.5 panels 2–3), returning to a depiction of a human element that more clearly aligns with the human topics of Francisco's words. With these overlapping trajectories of ebbing-and-flowing between iconicity and realism, and of zooming-in-and-out from human to nonhuman, this sequence becomes a nearly perfect visual palindrome that pivots on the passage's page turn, its various stylizations and layout almost identical whether read forward or backward. The passage's palindromic structure manifests a palpable sensation of back-and-forth, expanding and retreating, that heightens the wobbling and trembling of the human/nonhuman boundary the overall passage posits.

Interestingly, this boomeranging sensation is heightened as the boundary of the page also begins to falter under the pressure of two visual processes. First, the upward angle of Francisco's gaze in the second panel (Fig. 5.4 panel 2) violates point-of-view editing conventions and offers no clear sight line between him and the following frame's cow; the realization that the realistic panels were focalized from Francisco's perspective occurs after the fact, when the disembodied hand (Fig. 5.5. panel 3) is seen from the same high angle with which the protagonist gazes at his own hands a second later (Fig. 5.5 panel 4). Second, the continuity between the zoom-in on the fly (Fig. 5.4 panels 3–5) and the panel of ragged lines (Fig. 5.5 panel 1) is undermined when the shot zooms out, as it becomes unclear if this extreme close-up depicts the fly's wings, as assumed, or Francisco's hand. These dual processes of retroactive focalization and overextended zoom that simultaneously flow backward and forward over the page turn confront the reader with the adaptation's textual environment. They must physically turn the page back and forth, replicating and embodying the passage's boomeranging effect to untangle the palindrome's obfuscation of human and nonhuman.

Fig. 5.4 Francisco contemplates his son's operation

Moreover, it must be noted how this palindrome breaks Pulido's own compositional practices and principles, attesting to the uniqueness of this specific mise-en-page decision, even in a work that is "full of personal experimentation" (Jiménez). For Pulido, the two-page spread is the minimum unit of meaning, essential in planning and drafting *Nela* ("El proceso"). As Willem confirms, this process results in double-page spreads limited to "one narrative or symbolic idea," ending with a "hook that 'obligates' the reader to continue to the next unit" (83, translated by Willem). Here, however, Pulido allows the palindrome's play to flow onto the next page, while also making the "hook" link into a continuation of the present moment rather than a new one. This present moment will

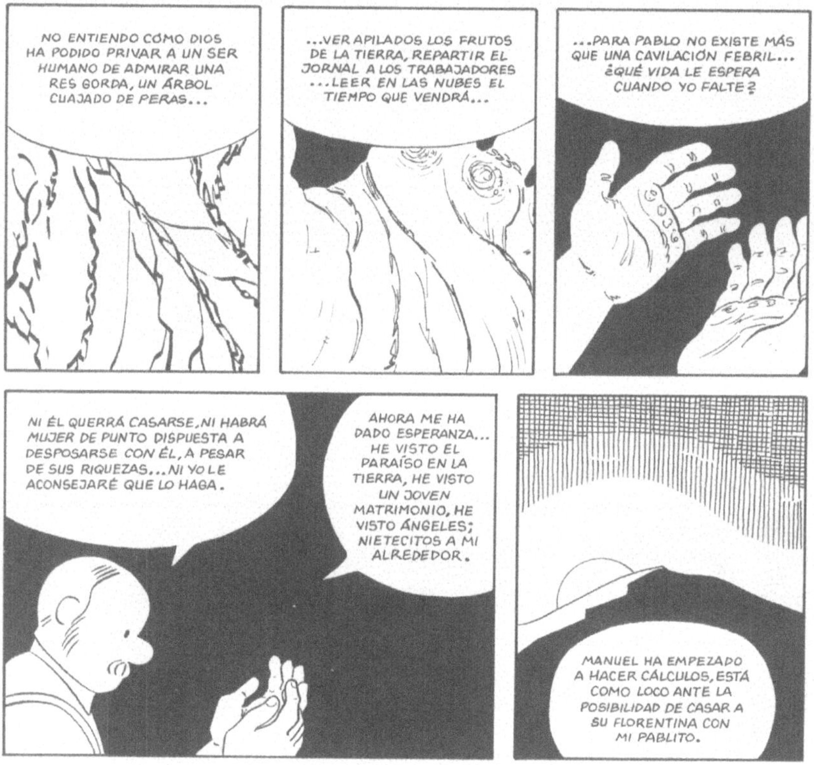

Fig. 5.5 The contemplation continues

only move forward, ironically, as the reversing, back end of the palindrome is read and sends the reader back to the previous page, literally throwing up a roadblock to the "onward, ever onward" of the novel's narration.

While these moments of ecological relationality between characters and their geo- and biospheres are subtle, the visual devices used to create them converge when the novel makes a textually explicit deviation from Galdós's original. As Teodoro and Pablo emerge from the mines, the latter explains that the sound emanating from the chasm into which Nela's mother threw herself to commit suicide is a "stream of air that escapes from the Earth's bowels, like a whistle, and it strikes against a torrent of water" (84). *Nela*'s adaptation of this passage hews quite closely to Pablo's explanation, but notably, the chasm's whistle is rendered as "*a moan* from the Earth's bowels, *a protest of her elements being stolen*" (Fig. 5.6, my emphasis). Remarkably, this textual modification occurs alongside an illustration that depicts the chasm with the same sound-symbolism previously given to Nela's singing: the wavy tentacles of sound now emerge from an opening in the Earth in an eye-popping gold, a color Pulido strategically uses in the novel to represent invisible (Willem 85), nonexistent (Willem 91), or fantastic (Ayala 241) things—in other words, decidedly nonhuman things. In short, as *Nela* explicitly condemns human intervention in the environment through words rather than the visual language Pulido prefers, his textual modification is still bolstered by the same visual tactics used in the novel's more subtle ecologically relational scenes.

With this more direct protest against extractive human practices, Pulido's adaptation clearly manifests "how human attitudes towards the environment... are adapted through time, across cultures, between texts" (Geal 6), given that *Nela*'s words stand in contrast to the ecological attitudes sustaining *Marianela*. To be sure, Galdós was enmeshed in the reality of emerging industrialization in Spain, with all the "antagonism between humans and their environments" (Estok 2) that would imply; his ecophobic understanding of human-nature "relationships that are very detrimental to our environment" (8), then, is not totally surprising. And though he was versed in Spanish schools of thought that promoted contact with nature and he was aware of, or at least sensitive to, early nineteenth-century ecological thinking, historically, emerging environmental consciousness would not consolidate in Spain until the conservationist movement of the 1910s, closer to Galdós' death than the publication of *Marianela*. And Galdós is not the only one who fails to critically engage with the deleterious environmental assumptions underpinning his era's thinking: while recent criticism has specifically examined the role environmental destruction plays in *Marianela* (Anderson; Bly; Sierra, "Confronting" and "Time"), a large majority of studies on nature in Galdós tend to be biophilic, focusing on the return-to-nature or nature-as-regeneration tropes at work in Galdós' broader representations of nature (Correa; Pageaux; Ruíz Pérez; Varey).

Fig. 5.6 "A moan from Earth's bowels"

The shift between the two respective eras' attitudes toward the environment becomes more pronounced if the metaphorical language each version uses to describe human interventions in nature is considered. On one hand, *Nela*'s ecological denunciation harkens back to pre-industrial metaphors that postulated that "minerals and metals ripened in the uterus of the Earth Mother, mines were compared to her vagina, and metallurgy was the human hastening of the birth of the living metal in the artificial womb of the furnace—an abortion of the metal's natural growth cycle before its time" (Merchant 11). The similar vocabulary in both Merchant's analysis and Pulido's denunciation is tellingly similar: the entrails and bowels, abortion and robbery. On the other hand, when Galdós's narrator returns to the mineshaft in daylight rather than in the darkness of Teodoro's initial journey, his descriptions become nearly epic (115–119), and he admiringly describes the human/machine amalgamation at work in the mines' chaotic and deafening madness. Galdós' description, ironically, praises much of the same equipment that Merchant includes in her condemnatory catalogue of the machinery that developed alongside the era's metaphors of "mastery and domination of nature" (11): the screws, the conveyor belts, the water pumps, and the tree stripers all make an appearance in two almost identical lists which are put to diametrically opposed ends: one admiring, the other decrying.

It must be underlined that Galdós's laudatory description of industry at the cost of the environment is not made by any particular character but rather by the narrator, an entity that Pulido entirely eliminates from *Nela* in what he describes as "one of the most radical decisions I made at the beginning of the process" ("Galdós, Unamuno"). While indeed drastic, Pulido's excision of the narrative voice, in fact, responds to the visual primacy he sees in the language of comics and which underpins his creative processes; eliminating the main source of "the weight of the novel's nineteenth-century rhetoric" (Willem 79), then, straightforwardly aligns with Pulido's underlying impulse that "everything that can be solved in silence should be solved in silence" (Ayala 240). Incidentally, however, this excision not only erases the narrative voice but also erases the environmentally harmful ideology embedded in it, and the elimination of both narrator and ecophobic attitudes is attested to by the fact that the chapter where Galdós makes his "epic hymn to industry and geological time periods" (Pulido, "Reocín") was rendered for *Nela* yet ultimately excised from the published version.[4]

Unraveling whether Pulido's decision to remove the narrator forms part of an un-, sub-, or conscious strategy to also remove "ecologically harmful assumptions" (Geal 5) from the source material as he creates *Nela*, or if *Nela* is opened to depicting ecological relationality precisely due to its "narrator-lessness" would require more exploration than this chapter permits. However, what this reading of *Nela* does confirm is the power of hybrid forms of storytelling that emphasize our interconnected position within the increasingly calamitous climatic conditions of the Anthropocene, as *Nela* not only adapts Galdós's novel to the new environmental conditions, constraints, and possibilities of a new ecological niche—graphic narrative—but it also uses that ecological transposition to illuminate and denounce the source material's environmental ideology, ultimately modifying it to align with contemporary environmental attitudes.

Furthermore, this analysis has shown how an allegiance to comic's visuality in the process of creating *Nela* not only buttresses the work's

[4] The full excised scene can be viewed on the post "Trabajo. Paisaje. Figura (2)" on Pulido's blog.

textual indictment of the source material's perpetuation of human's presumed superiority over nature, but it also allows *Nela* to use specific elements of the visual language of graphic narrative—sound-symbolic words, parallel word-image combinations, and mise-en-page decisions—to create moments of ecological relationality between his characters and their bio- and geospheres, which, upon being embodied by the reader, stand as a small corrective to the source material's ecophobia. This prioritization of image over word allows Pulido's hybrid form to "replicate without repeating" (Hutcheon 7), to copy with change (Bortolotti and Hutcheon 447) as he naturally and aptly adds ecological relationality to his work. While it remains true that such narrative performances cannot "give us the answer to how we can tackle the crisis" (Esben Svelstad 31) of our ever more precarious relationship with the environment, *Nela,* as a work in its own right, shows that hybrid graphic narrative might allow us to embody the kind of ecological recalibration our sense of self needs to bring it into a healthier relationality with nature; as an adaptation it demonstrates how we might retell our old stories without repeating their participation in the patterning of harmful human-nature relationships whose consequences we have no other choice but to now confront face on.

References

Anderson, Andrew. "Necessary Sacrifices: From Romanticism to Naturalism in Galdós's *Marianela.*" *Bulletin of Spanish Studies*, vol. 92, no. 6, 2015, pp. 907–929. *Taylor and Francis Online*, https://doi.org/10.1080/14753820.2014.947849.

Ayala, Joaquín. "Diálogo con Rayco Pulido." *Galdós y cuenta nueva*, edited by Carlos Brito Díaz and Isabel Castells Molina, Grupo Editorial Iberoamericano, 2023, pp. 235–248. *Dialnet*, www.dialnet.unirioja.es/servlet/libro?codigo=950543. Accessed 7 Aug. 2025.

Behiels, Lieve. "Luz y Oscuridad en *Marianela* de Galdós (1878) y *Nela* de Rayco Pulido (2013)". *XI Congreso Internacional Galdosiano: La hora de*

Galdós, edited by Yolanda Arencibia et al., Cabildo de Gran Canaria, 2019, pp. 766–82. *Actas del Congreso Internacional de Estudios Galdosianos*, https://revistas.grancanaria.com/index.php/cig/article/view/10344. Accessed 7 Aug. 2025.

Bortolotti, Gary, and Linda Hutcheon. "On the Origins of Adaptations." *New Literary History*, vol. 38, 2007, pp. 443–458.

Bly, Peter A. "Egotism and Charity in Marianela." *Anales Galdosianos*, vol. 7, 1972, pp. 49–65. *Biblioteca Virtual Miguel de Cervantes*, www.cervantesvirtual.com/nd/ark:/59851/bmcw09g1. Accessed 7 Aug. 2025.

Correa, Gustavo. "La Presencia de la Naturaleza en las Novelas de Pérez Galdós." *Thesaurus*, vol. 17, no. 3, 1963, pp. 646–65. *Centro Virtual Cervantes*, www.cvc.cervantes.es/lengua/thesaurus/pdf/18/TH_18_001_022_0.pdf. Accessed 7 Aug. 2025.

Dey, Subir, and Prasad Bokil. "The Narrative Functions of Sound-Symbolic Words in Comics and Graphic Novels." *Journal of Graphic Novels and Comics*, vol. 12, no. 5, 2021, pp.1080–1101. *Taylor and Francis Online*, https://doi.org/10.1080/21504857.2020.1853190.

Dobrin, Sidney I. "EcoComix: An Introduction." *EcoComix: Essays on the Environment in Comics and Graphic Novels*. McFarland, 2020.

Esben Svelstad, Per. "The Graphic Novel as Mediation of the Anthropocene: Allegory, Ignorability and Pedagogy in Javi Rey's Adaptation of Henrik Ibsen's *An Enemy of the People*." *European Comic Art*, vol. 17, no. 1, 2024, pp. 7–31. *Berghahn Journals*, https://doi.org/10.3167/eca.2024.170102.

Estok, Simon. *The Ecophobia Hypothesis*. Routledge, 2018.

Geal, Robert. "Towards an Ecocritical Adaptation Studies." *Adaptation*, vol. 6, no. 1, 2023, pp. 1–12. *Oxford Academic*, doi.org/https://doi.org/10.1093/adaptation/apad001.

Ghosal, Anindita, and Arindam Modak. "Imag(in)ing Decolonial Ecology: Exploring Tropical Eco-Graphic Narratives." *Etropic: Electronic Journal of Studies in the Tropics*, vol. 22, no. 1, 2023, pp. 175–96. *James Cook University Journals*, https://doi.org/10.25120/etropic.22.1.2023.3975.

Ghosh, Amitav. *The Great Derangement: Climate Change and the Unthinkable*. U of Chicago P, 2017.

Guynes, Sean A. "Four-Color Sound: A Peircean Semiotics of Comic Book Onomatopoeia." *Public Journal of Semiotics*, vol. 6, no. 1, 2014, pp. 58–72. *Public Journal of Semiotics*, https://doi.org/10.37693/pjos.2014.6.11916.

Heise, Ursula K. "The Hitchhiker's Guide to Ecocriticism." *Ecocriticism: The Essential Reader*, edited by Ken Hiltner, Routledge, 2015, pp. 164–77.

Hutcheon, Linda. *A Theory of Adaptation*. Routledge, 2006.

Khordoc, Catherine. "The Comic Book's Soundtrack: Visual Sound Effects in Asterix." *The Language of Comics: Word and Image*, edited by Robin Varnum and Christina T. Gibbons, U of Mississippi P, 2007, pp. 156–173.

Kukkonen, Karin. *Studying Comics and Graphic Novels*. Wiley Blackwell, 2013.

Jiménez, Jesús. "Rayco Pulido lleva por primera vez a Galdós al cómic con 'Nela'." *El País*, 5 March 2013, www.rtve.es/noticias/20130305/rayco-pulido-lleva-por-primera-vez-a-galdos-comic-con-nela/612641.shtml. Accessed 12 Nov. 2024.

Jones, Willam B., Jr. "*Classics Illustrated* and the Evolving Art of Comic-Book Literary Adaptation." *Oxford Handbook of Adaptation Studies*, edited by Thomas Leitch, Oxford UP, 2017, pp. 214–35.

McCloud, Scott. *Understanding Comics: The Invisible Art*. Harper, 1994.

Merchant, Carolyn. "Nature as Female." *Ecocriticism: The Essential Reader*, edited by Ken Hiltner, Routledge, 2015, pp. 10–34.

Pageaux, Daniel Henri. "Sur quelques 'rotours a la nature' dans la littérature espagnole fin de siècle'." *Romantisme*, vol. 30, 1980, pp. 49–59. *Persée*, www.persee.fr/doc/roman_0048-8593_1980_num_10_30_5420. Accessed 7 Aug. 2025.

Pérez Galdós, Benito. *Marianela,* edited by María Lucia Sotelo Vázquez, Penguin Random House, 2016.

Pulido Rodríguez, Rayco. *Nela: Una Adaptación de la Novela MARIANELA de Benito Pérez Galdós*. Astiberri, 2013.

Pulido Rodríguez, Rayco. "El proceso." *Nunca Trabajes Solo: Un Blog que parecía tener fecha de caducidad*, 23 March 2013, www.nuncatrabajessolo.blogspot.com/2013/03/el-proceso.html. Accessed 17 Nov. 2024.

Pulido Rodríguez, Rayco. "Galdós, Unamuno." *Nunca Trabajes Solo,* 8 March 2013, www.nuncatrabajessolo.blogspot.com/2013/03/galdos-unamuno.html. Accessed 25 Sept. 2024.

Pulido Rodríguez, Rayco. "Mi panorama." *Nunca Trabajes Solo,* 1 June 2013, www.nuncatrabajessolo.blogspot.com/2013/06/mi-panorama.html. Accessed 2 Dec. 2024.

Pulido Rodríguez, Rayco. "Reocín." *Nunca Trabajes Solo,* 19 April 2013, www.nuncatrabajessolo.blogspot.com/2013/04/reocin.html. Accessed 24 Nov. 2024.

Pulido Rodríguez, Rayco. "Se busca estilo." *Nunca Trabajes Solo,* 3 March 2013, www.nuncatrabajessolo.blogspot.com/2013/03/se-busca-estilo.html. Accessed 3 Dec. 2024.

Pulido Rodríguez, Rayco. "Trabajo. Paisaje. Figura (2)." *Nunca Trabajes Solo*, 17 Mar. 2013. www.nuncatrabajessolo.blogspot.com/2013/03/trabajopaisajefigura-2.html. Accessed 2 June 2024.

Pursall, Dona. "Growing Stronger Together: Representations of Active Eco-Citizenship within Contemporary Comics." *Closure: Kieler E-journal für Comicforschung*, vol. 7, no. 1, 2020, pp. 43–64, www.closure.uni-kiel.de/closure7/pursall. Accessed 30 Dec. 2024.

Ruiz Pérez, Ángel. "Visión Bucólica y Regeneracionismo de Galdós." *VIII Congreso Internacional Galdosiano (2005): Galdós y el siglo XX*, edited by Yolanda Arencibia et al. Ediciones del Cabildo de Gran Canaria, 2009, pp. 123–138.

Shepard, Paul. "Ecology and Man: A Viewpoint." *Ecocriticism: The Essential Reader*, edited by Ken Hiltner, Routledge, 2015, pp. 62–69.

Sierra, Sara. "Confronting Earth's History: Deep Time in *Marianela*, *Doña Perfecta*, and *La vuelta al mundo en la 'Numancia'*." *Anales Galdosianos*, vol. 50, 2020, pp. 55–67. *Project Muse*, https://doi.org/10.1353/ang.2020.0001.

Sierra, Sara. "Time and the Environment: 'Slow Violence' in Benito Pérez Galdós's *Doña Perfecta* and *Marianela*". *Confluencia*, vol. 35, no. 1, 2019, pp. 41–56.

Varey, John E. "Man and Nature in Galdós' *Halma*." *Anales Galdosianos*, vol. 13, 1978, pp. 59–73. *Biblioteca Virtual Miguel de Cervantes*, www.cervantesvirtual.com/nd/ark:/59851/bmcks729. Accessed 7 Aug. 2025.

Willem, Linda. *Adapting Spanish Classics for the New Millenium: The 19th-Century Novel Mediated*. Palgrave Macmillan, 2022.

Wrightson, Kendall. "An Introduction to Acoustic Ecology Soundscapes." *Soundscape: The Journal of Acoustic Ecology*, vol. 1, 2000, pp. 10–13. *Academia.edu*, www.academia.edu/4410713/An_Introduction_to_Acoustic_Ecology. Accessed 7 Aug. 2025.

6

"I Can Hear the Glaciers Melting": Tom Stoppard's Radio Drama *Darkside* and the Exploration of Ecological Ethics via Pink Floyd's *The Dark Side of the Moon*

Thomas Grochowski

Radio is substantially more efficient, ecologically, than television and internet-based forms of communication, or even audio podcasts, which use less energy than audiovisual media but still rely on network servers for cloud storage. Radio is an example of what Jacob Smith calls "eco-sonic media." Its "comparatively small energy requirements" (110) and low production costs can make it play an important role in a more ecologically sound (as it were) media landscape. "Radio," according to Smith, "should have a central place in a green-media agenda that aims to provide content to the largest number people at the lowest cost to the planet's resources" (111). Smith's *Eco-Sonic Media*, an impressive work of media archaeology, offers an "historical approach to greening the media" (5), providing several historical case studies, beginning with his discussion of the initial use by the phonograph industry of biodegradable shellac through its transition to the more toxic polyvinyl so familiar to record-buyers of my generation, and ending with a discussion of radio drama

T. Grochowski (✉)
St. Joseph's University, New York, NY, USA
e-mail: tgrochowski@sjny.edu

P. Demory (ed.), *Ecoadaptation*, Palgrave Studies in Adaptation and Visual Culture,
https://doi.org/10.1007/978-3-032-09847-4_6

narratives "organized around spatial as well as temporal 'Ends of the Earth'" (113). Like its visual precursor, cinema, and its successor, television, radio has often been used as a forum for narratives adapted from other media, notably literature but also film—and in the most famous American radio drama broadcast, a literary adaptation that employed the techniques of radio itself, Orson Welles' "War of the Worlds" broadcast of 1938. Smith examines a lesser-known Welles adaptation as part of his effort to "foster a dialogue [in Bakhtinian and dramatic senses] between radio plays, American environmental literature, and eco-criticism" (112). As someone long influenced by Bakhtin and the work of those who have applied his theories to cultural critique, I offer this chapter as my own contribution to the dialogue, discussing a twenty-first-century radio drama adaptation that embodies many of the qualities of eco-sonic media that Smith champions.

Tom Stoppard's *Darkside: A Play for Radio Incorporating* The Dark Side of the Moon *by Pink Floyd* (2013, hereafter referred to as *Darkside*), merits considerable attention as illustration of the kind of radio drama of interest to Smith, a narrative about the relationship between ecological crises (the "spatial ends of the earth") and madness, and about the role philosophy can (or cannot) play in understanding that relationship. Stoppard, who had written several plays for radio prior to and after his breakthrough 1966 stage play *Rosencrantz and Guildenstern are Dead*, was approached by the BBC to commemorate the album's fortieth anniversary, though when interviewed about the play he acknowledged that a friend gave Stoppard a copy of *Dark Side of the Moon* when it was released in 1973, telling him that he "must" write a play about it (Rohter C5). Stoppard's deployment of the band's most famous album expresses his familiar themes concerning the role of philosophy in our world; it also makes a statement about the contemporary ecological crisis, which is of chief concern for its protagonist. As an adaptation of Pink Floyd's album, *Darkside* extends, or, to use a Bakhtinian term favored by Smith, *reaccentuates*, the record's existential themes, repositioning its music and especially its *sounds* into a related narrative.

Darkside's narrative centers around the mental state of the protagonist Emily McCoy (voiced by Amaka Okafor), a young woman with a mental illness that eventually will be treated by brain surgery. McCoy is enrolled in a philosophy course entitled "Great Thinkers," and the concepts from the course become the main field of play in the drama, in the form of thought experiments. The play's fourteen scenes move from diegetic spaces that may be differentiated between "the real" (the lecture room, the hospital) and less defined locations that form what I am calling a thoughtscape, since they seem to be generated through McCoy's own thoughts. Throughout, the album plays either as background underneath the drama's dialogue or on its own, as thematic commentary in relation to the drama. (For nearly twenty minutes of this fifty-four-minute drama, all one hears is the album content.) The play thus "curates" (in Neil Verma's words) three formats—radio drama, concept album, thought experiment—in such a manner as to demonstrate the "resonant contact" (191) each has with the other two.

When considering *Darkside*'s incorporation not just of the album but of thought experiments, we can see more clearly the play's function as a dialogical ecoadaptation; as Bryce Ezell (72) puts it, Stoppard's use of the thought experiment illustrates that philosophers, in the practice of constructing thought experiments, engage in creating drama. Indeed, for Ezell, the thought experiment *must* become drama for it to become embodied into a concrete form to connect to an audience. The radio play is an ideal format for this revelation, since it creates an embodiment of the abstract concepts through voice, sound effects, and music, an embodiment that is different from that of live stage plays. Radio drama also is potentially "a powerful tool for conveying the dangers of global warming," due to its ability to create drama from the unseen, like (until very recently) the impact of climate change (Smith 165–166). The play's "dark ecology" (Smith 110–141) becomes a fertile ground for the creation of connections between Pink Floyd's album and the function of moral philosophy. In addition to a shifting between abstract philosophical experiments and the concrete imperative of "glaciers melting" (a phrase repeated in the play as a shorthand for ecological crisis), *Darkside* also shifts between the abstract and the concrete because of its form: through its dialogue, music, and sound effects, radio attempts to make material the characters and scenes in the drama in the imagination of the listener. At the same time, because it is implied that McCoy's travels through the thoughtscape are a function of *her* imagination, the material existence of the thought-experiment characters she encounters cannot be established in the way that her instructor Mr. Baggott's and his classroom's can be. I argue that *Darkside* is an ecoadaptation in form (as a radio drama) and in content, because the protagonist works through and challenges the thought experiments posed, by identifying the choices created by those experiments as either irrelevant or even exacerbating the ecological crises facing humans.

Radio adaptations provide engaging forms of reaccentuation; not only can written speech in fiction become vocal speech, but also descriptions of landscapes can be sonically recreated, as Smith observes in discussing the *Mercury Theatre of the Air*'s production "Hell on Ice." The Orson Welles production is an adaptation of a nonfiction book about a failed 1879 North Pole expedition, one that works extensively with the journals of the expedition's leader, Captain George Washington DeLong, and which Smith describes as a "radiogenic book, with vivid descriptions of sounds" of the landscape (115). Smith's analysis of the broadcast draws parallels to a signal work of American ecological literature, Thoreau's *Walden*. Using Bakhtin's concept of the literary chronotope, Smith argues that, like *Walden*, "Hell on Ice" begins with an adventure novel chronotope that is usurped by the idyllic chronotope: both begin with a story of individuals setting out on an adventure in nature, but as the individuals become stationary (in the case of *Walden*, by choice; in the case of "Hell on Ice," by circumstance), the narratives become more an accounting of the daily lives of the central figures and the descriptions of the natural space around them (116–117). Smith's purpose in placing *Walden* into dialogue with "Hell on Ice" is to "reorient... the matrix of texts in which 'Hell on Ice' is situated," allowing the listener "to hear Welles' broadcast in new ways" (116). Also added to this "matrix of texts" is the writing of Sierra Club founder John Muir, who accompanied the first expedition that attempted to find out what happened to Captain DeLong and the crew of his ship, the *Jeannette*. Smith "hears" Muir's description of the formidable, natural world of the Arctic in the modernist sound reproductions of that space. Smith notes the irony of the program's sense of the natural world as being formidable, towering over humanity's efforts to explore it; given the rate at which the polar caps are melting in the twenty-first century, it is impossible to listen to this production without the awareness of how temporally bound all life on the planet is. Through his rehearing of "Hell on Ice," Smith thus argues that the broadcast offers not only a social critique of America's full entry into industrialization and empire (a common motif in Welles' own work, notably in his radio and cinematic adaptations of Booth Tarkington's *The Magnificent Ambersons*), but an ecological one as well.

Where Welles' radio drama provides sounds of the Arctic landscape as powerful and as overwhelming the humans in their endeavors to explore it, Stoppard's *Darkside* uses the repeated phrase "glaciers melting" to emphasize that landscape's vulnerability in the face of the industrialization Welles (and his immediate sources) warned against. Following Smith's Bakhtinian approach to "radio's dark ecology," I see, or rather, "hear," *Darkside* reaccentuating *The Dark Side of the Moon* from an atmospheric existential concept album into a multivocal narrative of personal and social madness, with the latter being couched in planetary crisis (or, to use Smith's phase, the end/s of the Earth).

As a radio play, *Darkside* is unique in the realm of adaptations of popular music. As Ian Inglis has noted, popular music has become a source for cinema, with "the transformation of vinyl to celluloid" proceeding "along several trajectories" (314). These include adaptations of narrative songs such as "Alice's Restaurant" and "Harper Valley P.T.A."; full narrative "rock operas" like The Who's *Tommy*; or concept albums like *Sgt. Pepper's Lonely Heart's Club Band*. The BBC's "heritage" adaptations for radio have been in the field of literature; only recently has the Corporation broadcast a narrative rock opera adaptation, that of the Kinks' *Arthur* in 2019. *The Dark Side of the Moon* is not a narrative album like *Arthur* or *Tommy*. (Considering Smith's interpretation of "Hell on Ice," we can note that the album's lack of narrative makes it akin to both the nonfiction *Hell on Ice* and *Walden*.) Often characterized as an exemplar of "space rock," a subgenre the group had been associated with since its formation in 1967 (Romao 211), it contains several instrumental tracks, and the relatively sparse lyrics are meditative; they don't tell stories even within individual songs. Rather, it is a concept album linked via a set of existentialist themes centered on everyday concerns: aging, death, madness, greed, and grappling with life's meaning. As Lawrence English has noted, the album evokes cultural touchstones such as Jungian theories of archetypes, Blake's prophetic works, and Zen Buddhist's koans, thus aligning it with Bakhtin's dialogic theories of literature. (It is also a literalization of polyphony, with two different lead singers, the wordless vocal performance by Clare Torry, and the different voices heard in snippets throughout the album.) The themes on the album are "themes that frame our lives, [showing] us that we choose whether that frame is a box around us or a window to be climbed through" (4). With sound effects such as heartbeats, chiming clocks, cash registers, airplane crashes, and the human voices discussing being drunk and going mad, *The Dark Side of the Moon* creates "its own universe" (Easlea 309), described by K. J. Donnelly as a "soundtrack LP without a film, constructing something that approximates a cinematic experience within domestic space" (46).

In reaccentuating *The Dark Side of the Moon, Darkside* emphasizes the existentialist themes of the album's lyrics, themes that, according to David Detmer, reflect Roger Waters' own philosophical views (as expressed in interviews as well as his lyrics), especially as expressed by Camus, Sartre, and Beauvoir. Waters' words suggest an awareness of "the fact that we... are already immersed in a world full of economic, political, and cultural structures that we have not chosen" and of "the inescapability of our responsibility for how we respond to those structures" (66). Thus, *The Dark Side of the Moon*, perhaps more clearly than any other Floyd album, presents us with a sense of ethical obligation. While alienation is a central theme of almost all of Pink Floyd's body of work (Macan), *Dark Side of the Moon*'s tone remains definitive in its lack of sanctimony associated with later Floyd records like *The Wall* or *The Final Cut*. In an interview conducted in 2003, Waters spoke of the album thus: "If there is any central message it's this: this is not a rehearsal" (Clerk, qtd. in Easlea 309). A near-identical statement will be spoken by a character in the play; the "Wise One" gives the protagonist this advice: "The secret of life is: this is not a drill" (Stoppard 19).

The existential themes of *Darkside* were a natural fit for Stoppard, whose other works have become famous for their witty critique of abstract philosophical discourses. Stoppard himself drew a connection between *Darkside* and his most famous play, *Rosencrantz and Guildenstern are Dead*, because both required working within a specific set of parameters. With *Rosencrantz*, "I had to work inside Shakespeare's play. Working inside Pink Floyd's album is the same syndrome" (Rohter C5). *Rosencrantz* contains the entirety of the two minor characters' appearances in *Hamlet*; *Darkside* utilizes the entirety of *The Dark Side of the Moon*. Heebon Park-Finch, borrowing from Gerard Genette, describes *Rosencrantz* as a "transfocalization" of Shakespeare's play. Where Genette uses the concept in more hypothetical (if playful) terms (305–313), Park-Finch applies it more forcefully, arguing that Stoppard "offers an alternative perspective as he re-writes, re-views, re-activates and re-configures" (185) the source play. The story told in *Hamlet* remains "intact" but reimagined through the perspective of the two minor characters, philosophizing on the nature of existence much as the prince constantly does, though with even less control over their fate. Where Park-Finch uses the visual term *transfocalization* to demonstrate how Stoppard constructs an "alternative space" (Park-Finch 187) with *Rosencrantz*, signaled in the script via lighting changes and other shifts of *mise-en-scène*, it would seem more appropriate to follow Smith's use of Bakhtin while discussing *Darkside*, to say that Stoppard *reaccentuates* the sonic and lyric themes of *The Dark Side of the Moon* to address philosophical questions that he has often raised in other works. Where other critics have suggested that *Rosencrantz* decenters Shakespeare's play (Sanders, qtd. in Park-Finch 192), *Darkside* decenters philosophical inquiry itself, questioning its relative usefulness in the context of impending climate-related doom.

The play's opening sets the pattern for the resonant contact the album, radio drama, and thought experiment share. After the BBC announcer's introduction of the play (included on Faber and Faber's release of the CD recording, packaged with the script), *Darkside* begins with... *The Dark Side of the Moon*. The play's opening is identical to the first seventy-four seconds of the album, the overture *musique concrète* piece "Speak to Me," which begins with a beating heart, to which are added the noted sound effects that appear throughout the album, and the voices that have become part of the album's lore, belonging to people associated with the band or with EMI Studios answering a series of questions written by Roger Waters on cards. Most of what is heard in "Speak to Me" is that of roadie Peter Adamson answering the question, "Do you think you're going mad?" Among his heard answers are "I've been mad for fucking years" and "very hard to explain why you're mad, even if you're not mad." This beginning, which establishes a tone for the album, is reaccentuated by Stoppard by applying it to themes beyond mental illness. Emily McCoy's statement about "hearing voices... talking and laughing about being mad" (Stoppard 14) is the *album listener*'s experience, connecting the *radio drama listener* to the protagonist. Near the end of the album's opening track, a portion of Clare Torry's famous vocal from "The Great Gig in the Sky" is heard, and in *Darkside*, the sound of a speeding train is layered into the soundtrack. The train sound effect certainly would not be out of place in the original album, but in fact it marks the beginning of the dramatization of the first thought experiment, that of the trolley problem, famously posed by philosopher Phillipa Foot in the 1960s.

Structured by Foot as an analogy illustrating the problem of determining whether one should sacrifice one life or five lives (by pulling or not pulling on a railway switch), the trolley problem is adapted by Stoppard into a radio superhero drama, done with voices speaking in "American cartoon style" (Stoppard 7). Unlike most American radio adventure stories like *Superman*, this minidrama ends tragically, with a boy killed at the expense of the others on the train. At the same time, the scene is filled with incongruities that can be called both comical and absurd. The hero of this drama is "Ethics Man," a moral philosopher. The exaggerated American voices parody the style of similar radio adventure programs from the thirties. The drama transitions to the instrumental opening to the album's second song, "Breathe," and the second scene, set in the interior of the lecture hall, begins, with the instructor, Mr. Baggott (voiced by Rufus Sewell, who also voices Ethics Man) asking a class if they think Ethics Man did the right or wrong thing (asking via a show of hands, a clever joke for a radio program). McCoy, who did not raise her hand, asks, "Who was on the train? Who was the boy who got hit by the train?" (Stoppard 9). These questions are dismissed by Baggott, since for him, the dramatized trolley problem is an illustration of the philosophical perspective he is explaining to the class, that of utilitarianism. McCoy's questions push the thought experiment into the discourses of everyday life, even as they send her further into the thoughtscape, where she encounters characters who are thought experiments, beginning with the "Boy" (voiced by Iwan Rheon) who was killed by the train. This opening establishes the pattern through which the protagonist will cut the Gordian knots of thought experiments and direct the other characters—"real" and "thought experiment" alike—toward the ecological crisis. While McCoy admits to hearing voices in the context of her diagnosis, she also says (after the surgery) that she can "hear the glaciers melting" (Stoppard 47), another way that Stoppard shifts language from the abstract to the concrete.

As McCoy travels with the Boy through the thoughtscape—much the way that Dorothy travels through Oz in *The Wizard of Oz* (Ezell 75)—she encounters several other thought experiments, variations on the trolley problem, along with an ecoethics problem posted by ecologist Garret Hard known as the tragedy of the commons, which describes how people behaving unethically while using common land led to turning the region (dramatized by Stoppard, a valley) into a wasteland—an obvious metaphor for the global climate crisis. Told this tragedy by a thought-experiment character (voiced by Adrian Scarborough) simply known as the Fat Man[1] (who was sacrificed in a leaky hot air balloon experiment to save the lives of three thin balloonists), McCoy again cuts through the problem by pointing out that had the people who shared common land in the thoughtscape valley decided *not* to cheat, the valley could have sustained itself. The dramatization of these thought experiments, and of McCoy's persistent challenges to them, are placed into overlapping dialogue with the radio drama format and with *The Dark Side of the Moon*; the album's own tension between the abstract and the concrete (manifest in its use of sound effects of concrete objects like clocks and cash registers in contrast to the abstract ideas in the lyrics and in the nonlyric vocal by Torry) becomes reaccentuated into the play's narrative context, in two significant ways.

First, the album's instrumental passages are audible underneath the play's dramatic action. Scene Three begins with sounds of McCoy and

[1] The name "Fat Man" is not a mere descriptor; it refers to the character played by Sidney Greenstreet in *The Maltese Falcon*, and it sets the pattern for the character, who frequently speaks in movie quotes, beginning with a line from *The Wizard of Oz* ("Toto, I've a feeling we're not in Kansas anymore"). Stoppard's reference also alludes to one of the myths around the album, that there are correspondences between the record and the 1939 film (when one begins to play the album after the MGM lion's third roar). For more details, see Savage.

the Boy running in an undefined exterior, with explosion sound effects suggesting a violent, war-torn location, to which is added the instrumental track "On the Run," which features hypnotic synthesizers and drum cymbals playing at a fast pace over various airport announcements. Where the album implies the rushing of people through airports (a common space for rock musicians), *Darkside* reaccentuates "On the Run" as a background for the nervous movement of the two characters across a vaguely threatening landscape, as they discuss philosophy and the questionable ethics of Ethics Man. McCoy questions Baggott's dismissal of her classroom challenge by pointing out that in killing the Boy, Ethics Man may have killed "the one who stopped the glaciers melting" (11), the first specific mention of the play's ecocritical theme. It is the first reaccentuation of the ethical themes of the album's lyrics, taking general lines like "don't be afraid to care" (from "Breathe") and grounding them in the particulars of contemporary ecological concerns. The plane-crash sound that closes "On the Run" begins another turn in the narrative toward thought experiment/radio drama parody: as McCoy and the Boy observe someone in a parachute floating downward, an announcer in "American-cartoon-style" declares, "Is it a bird? Is it a plane? It's... *Übermensch*!" (15). The tagline comes from the Superman radio (and later, television) series, with the German translation of "Superman" shifting *Darkside* back toward philosophy, as Baggott/Ethics Man, the man in the parachute, explains that he, now a Nietzschean egoist, has survived the crash by having the will to take the one remaining parachute left (leaving a banker and a politician to die in the crash). Instrumental passages within songs that contain lyrics are also used as narrative soundtrack for both the thoughtscape and the real spaces: the opening instrumental to "Time" becomes backdrop for Baggott's class lecture on Nietzsche, and the segue between "Money and "Us and Them," featuring voices from the album, plays underneath the discussion after McCoy's surgery to remove "all the voices," a surgery that in fact "fails," since she still enters into dialogue with the Boy and the Fat Man afterward. McCoy thus represents the dialogic discourses favored by Bakhtin in contrast to the lectures and "mansplaining" of Baggott. Her rediscovery of the Boy near the end of the play is implicitly a favorable one, for "listening to the voices" allows McCoy to connect to the ecocrisis; her rejection of

Baggott's philosophical discourses is an expression of an empathy needed to solve the crisis.

Secondly, the lyric/vocal parts of the album's songs extend the philosophical discussions McCoy encounters and often challenges; she is often placed into dialogue with those lyrics, as is the case with "Time," whose vocals begin after the Wise One tells her the "this is not a drill" secret of life. After the first chorus, and over the guitar solo, McCoy expresses disappointment with learning that life's secret is "what your mother's been shouting up the stairs at you—'Wake up, Emily, your life is going by!'" (20). The play's use of the nonlyric vocal "the Great Gig in the Sky" reaccentuates the individualized grief expressed by Clare Torry's singular, famous vocalizing into a broader cry for help to "save the earth." Its position in the narrative, as in the album, is right in the middle: McCoy finds herself about to make a speech on how we can transform wasteland into garden, encouraged by the Boy's advice to hold her thoughts like breaths, then let the breath go, "soundwaves of pure thought" (27), again illustrating the dynamic between the abstract thought and the concreteness of sound. When she lets go of the breath, Torry's vocal part kicks in, and when it fades, McCoy delivers her speech (in verse) that suggests that the key to renewal is to return to human nature, a nature that is "born to kindness" (28), echoing again the lyric "don't be afraid to care." This altruism will be called into question when she is held on trial for witchcraft (in another dramatized thought experiment, that of the prisoner's dilemma), and the banker and politician return to speak against her "hippy-dippy, do-goody-goodery" (36), leading into the album's hit single "Money." A satire on greed in the album, in *Darkside* "Money" retains its satiric function as representative of the banker and politician's values, while also becoming soundtrack in its long instrumental passages, with its machine-gun-style guitar progression reaccentuated into the sound *of machine guns*; in the frantic narrative underneath the guitar and saxophone solos, McCoy and the Boy's execution (by burning) is thwarted by the people whom Ethics Man saved in the trolley car, and as the two run from the army, they find the valley is rapidly blooming, regrowing, returning to its old splendor, only to have the banker and politician begin to sell and organize the land to support their capitalist agenda.

As *Darkside* reaches its conclusion, McCoy presents a more apocalyptic perspective regarding the crisis, first to Baggott (visiting her at the hospital), where she describes her helplessness due to "Injustice and unfairness... running free like they own the earth" to the point where soon "the earth won't be worth owning" (50), and then as part of a monologue where she describes planetary evolution going back to the Big Bang Theory, followed by a speech forecasting planetary doom, while the album's "Brain Damage," a song about madness whose lyrics are used to refer to McCoy's surgery, plays underneath. Stoppard creates a dialogue between McCoy and one of the album's voices: after a short speech that begins "The ice is melting," a voice from the album is heard, saying, "I can't think of anything to say, really," followed by laughter, over which McCoy continues, saying, "We're dying of consumption" (52). Stoppard's conclusion is more ambivalent, as McCoy asks as part of her final speech, "Do you believe in the juggler?" Her final words are followed by the closing of the album, the beating heart that begins it. The "juggler" is the "juggler on the radio," a concept the Boy explains to Emily shortly after they meet:

> There's a juggler on the radio. He sounds exactly the same as if there's no juggler. There's lots of people listening to the radio, and some are saying, 'I believe in the juggler!' and some are saying, 'there *is* no juggler!" and then there's a few philosophy-type people who are asking, 'how is a juggler you can see, hear, smell or touch different from no juggler?' But there's nothing any of these people can tell each other about the existence or non-existence of the juggler. (13)

McCoy herself is a juggler, rejecting the idea that one must choose either the boy on the track or the people on the train. (In the current ecological crisis, as Verma points, out, humanity is both the boy *and* the people.) *Darkside* juggles the ideas and sounds of the album within the formats of the radio drama and thought experiment. For McCoy, the juggler becomes a metaphor on the precarious necessity of holding multiple contradictory thoughts at the same time; to believe in the juggler is to believe in the hope that humanity can connect to its "natural" state of kindness to prevent ecological devastation. In using the form of the radio drama, Stoppard has utilized an eco-sonic medium to mine the many voices, the "dark ecology" of *The Dark Side of the Moon,* into a contemporary ecocritical narrative, placing into action the abstract, ethical concerns present in Roger Waters' lyrics.

Works Cited

Detmer, David. "Dragged Down by the Stone: Pink Floyd, Alienation, and the Pressures of Life." *Pink Floyd and Philosophy*, edited by George A. Reisch, Open Court Publishing, 2007, pp. 61–80.

Donnelly, K. J. *Magical Musical Tour: Rock and Pop in Film Soundtracks.* Bloomsbury, 2015.

Easlea, Daryl. "'Cruising for a Bruising': How *The Dark Side of the Moon* Made Pink Floyd Famous Beyond Their Wildest Dreams and Instigated Their Downfall." *Routledge Handbook of Pink Floyd*, edited by Chris Hart and Simon Morrison, Routledge 2023, pp. 306–17.

English, Lawrence. *More Than Music:Cultural Stirrings from the Dark Side of the Moon.* Algora Publishing, 2021.

Ezell, Bryce. "Tom Stoppard and the *Darkside* of Drama, Theatre, and Philosophy. *Modern Drama*, vol. 63, no. 1, Spring 2020, pp. 63–82. *ProjectMUSE*, muse.jhu.edu/article/752498. Accessed 16 Aug. 2025.

Genette, Gerard. *Palimpsests: Literature in the Second Degree*. U. of Nebraska Press, 1997.

Inglis, Ian. "Music into Movies: The Film of the Song." *A Companion to Literature, Film, and Adaptation,* edited by Deborah Cartmell, Wiley-Blackwell, 2012, pp. 312–329.

Macan, David. "Theodor Adorno, Pink Floyd, and the Psychedelics of Alienation." *Pink Floyd and Philosophy*, edited by George A. Reisch. Open Court Publishing, 2007, pp. 95–120.

Park-Finch, Heebon. "Hypertextual Adaptation: Humanistic Enquiry through Transfocalization in Stoppard's *Rosencrantz and Guildenstern Are Dead*." *Journal of Adaptation in Film & Performance* vol. 5, no. 2, 2012, pp. 183–196. https://doi.org/10.1386/jafp.5.2.183_1.

Rohter, Larry. "An Author Dives into Pink Floyd." *New York Times,* 25 Nov 2013, pp. C1+.

Romao, Tico. "In Search of Space (Rock): Pink Floyd and Genre Formation in Rock Music." *Routledge Handbook of Pink Floyd*, edited by Chris Hart and Simon Morrison, Routledge, 2023, pp. 211–225.

Savage, Charlie. "Pink Floyd, The Wizard of Oz, and Me." *New York Times Magazine* online, 21 June 2023. www.nytimes.com/interactive/2023/06/21/magazine/dark-side-of-the-rainbow-pink-floyd-wizard-of-oz.html. Accessed 16 Aug. 2025.

Smith, Jacob. *Eco-Sonic Media*. U. of California Press, 2015.

Stoppard, Tom. *Darkside: A Play for Radio Incorporating* The Dark Side of the Moon *By Pink Floyd.* Faber and Faber, 2013.

Verma, Neil. "A Juggler on the Moon: How Sounds Think in Tom Stoppard's *Darkside*. *Recherches sémiotiques / Semiotic Inquiry*, vol. 36, nos. 1–2, 2016, pp. 181–197. https://doi.org/10.7202/1051184ar.

7

The Ark of the Moral Universe: Recycling Genesis Toward Environmental Justice

Matthew Page

If ecocritical adaptation studies can demonstrate "how human attitudes to various aspects of the nonhuman world around us change and adapt through time" (Geal, "Towards" 6), then one pivotal area for analysis is ancient texts which have been adapted multiple times over the centuries.[1] In *Literature for a Changing Planet*, Martin Puchner describes "large timescales as particularly important for understanding the relation between storytelling and the environment" (10). One such recent adaptation is Darren Aronofsky's *Noah* (2014), which reworks the biblical story from Genesis into a modern biblical epic. Aronofsky's film is a late addition to a long tradition of adapting the flood story, which has

[1] My use of "nonhuman" throughout should not be read as deliberately referencing a spiritual/divine realm but as an alternative way to refer to "the natural world," "the environment," or "creation."

M. Page (✉)
Loughborough, UK
e-mail: matthew.w.page@gmail.com

P. Demory (ed.), *Ecoadaptation*, Palgrave Studies in Adaptation and Visual Culture,
https://doi.org/10.1007/978-3-032-09847-4_7

inspired countless writers, artists, theologians and, more recently, film-makers, whose work has responded to existing interpretations and shaped the works that followed. This rich tradition enables us to trace how the story has reflected and informed human attitudes and behavior across many centuries and varying cultures and to track an evolving story as it mutates to adapt to different contexts.

Examining 4000 years of flood story adaptations ecocritically reveals much about humanity's interaction with the nonhuman world. Yet while we might hope a narrative's evolution becomes progressively ecophilic with time, this remains as uncertain for cross-medial adaptation as it is for biological evolution, where more complex organisms are not necessarily those which persist. Even when tied to a literary anchor like Genesis, genre, character, tone, and meaning vary, sometimes resulting in the same story being used in opposing ways.[2] Aronofsky's film demonstrates that adaptations can adopt more ecophilic approaches than their primary sources, but that this may be as much to ameliorate cultural anxieties as to persuade audiences to change. Moreover, while the film represents something of a high water mark in flood story narratives, its retention of ecophobic elements and the history of adaptations suggest this upward trajectory will not continue indefinitely.

The Story of the Flood

Flood narratives were ubiquitous within the ancient world (Gaster's 1969 study counted 300): flooding was a regular environmental hazard in antiquity and boats were a means of survival. The most widely known flood narrative in western culture is Genesis's story of Noah's Ark. Yet just as "history is itself always an adaptation of some earlier history" (Leitch 10), Genesis itself is composed of "recycled" materials. The story which today is associated with Noah originates from Sumerian and Akkadian

[2] For example, following the English Civil War, Marvell used the story in favor of the Parliamentarians. Dryden used it to support the Royalists.

myths written a thousand years earlier and, before that, to oral tradition.[3] In the *Sumerian Deluge* (the first remaining written version of the story from around the second millennium BCE) the flood hero is not Noah, but Ziusudra. The *Atrahasis Epic* (c. 1750 BCE, according to Finkel 104), similarly has "basic elements" that "closely resemble those of Ziusudra" (Collins 10), as does our "first great work of world literature," *The Gilgamesh Epic,* written a millennium later (Puchner 14).

Genesis shares the primary plot of these earlier texts: "the broken bond between the gods and humanity... restored after the deluge by a sacrifice of ark animals offered by the flood hero" (Stavrakopoulou 213). Yet the story's incorporation into the Jewish scriptures 2500 years ago, and its subsequent inclusion in the Christian Bible and the Muslim Qur'an, have resulted in reshaping and recycling ever since, from the violent details of people being flayed alive in boiling flood waters, (e.g., *Sanhedrin 108b* of the *Babylonian Talmud*, 500–600 CE) to more child-friendly variations in evidence since the seventeenth century when "commercially produced toy Arks and animals... began appearing" (Shepherd 6). Indeed, since the emergence of cinema, most dramatic visual adaptations have been produced with children in mind, even operas like Benjamin Britten's *Noye's Fludde* (1958). Such variations demonstrate how far such a literary vessel can drift if left untethered.

The earliest flood narratives, including Genesis itself, derive in part from the trauma resulting from major, or perhaps recent, floods. They (also) function as a means of dealing with widespread cultural anxieties or pressures, including environmental threats, often by offering reassurance or challenging audiences to adjust their behavior. Clusters of adaptations within a short time suggest the presence of specific anxieties. For example, four significant literary flood adaptations arise in England, over a period of just fifty years, during the seventeenth century, when civil war, plague, and flooding connected with the mini-ice age, converged with devastating consequences.[4] Not dissimilarly, the intertitles for Curtiz's

[3] Genesis itself comprises multiple sources one of which may have recycled one of the others (Carr 61).

[4] See Mottram for the wider context. Further afield, Joost van den Vondel's *Noah, of Ondergang der eerste weerelt* (1667) describes the floodwaters rising gradually well before the rain falls, forcing herders to migrate, a strikingly apt anticipation of rising sea levels today.

film *Noah's Ark* (1928) specifically compare the flood with the First World War, "a Deluge of Blood drowning a World of Hate."

Ecophilia in Flood Narratives

The most obvious, yet often overlooked, element of the story is the centrality of the nonhuman world: the animals, plants, and indeed the flood waters themselves. The story's major nonhuman element is the animals, whose presence has grown over time. In *The Sumerian Deluge*, the building and boarding of the ark is absent, and animals feature only as sacrifice victims. The later *Atrahasis Epic* introduces the idea of animals coming two by two, which Genesis expands to seven pairs for "clean" animals. Later Jewish texts increase the animals' importance. In *B. Sanhedrin 108b*, for example, Noah demonstrates "an intimate knowledge of and desire to learn from animals" (Lilly 40). As *visual* adaptations of the story develop, animals become even more prominent. The early fifteenth-century *Chester Mystery Plays* displayed painted representations of multiple animals (see Rhys 34). Paintings by Jacopo Bassano, Jan Brueghel the Elder, Giovanni Benedetto Castiglione, and Hieronymous Bosch position the animals prominently in the foreground. Most on-screen adaptations, from *The Tale of the Ark* (Cooper 1909) to *El arca* (Buscarini 2007) and *Noah's Ark: A Musical Adventure* (Leo and Machado 2024), even tell the story from the animals' point of view (as did Julian Barnes' novel *A History of the World in 10½ Chapters* (1987)). While this places animals at the center of the film, suggesting an ecophilic perspective, it also anthropomorphizes them, diminishing their nonhuman nature and contaminating them with human traits.

Initially, plants are represented only by the dove's olive branch. Yet plants also feature at the end of the Genesis story when Noah pioneers viticulture, suggesting either reunification with the nonhuman world, or further exploitation.[5] Later, *Genesis Rabbah* (c. 400 CE) features Noah bringing cuttings with him to plant post-flood, highlighting the

[5] Some scholars suggest this was Noah's original role in *Genesis*, with the flood-hero narrative assigned much later (Bailey 208).

devastation facing the plant kingdom (*Midrash Rabbah*, Gen. 31:14). Michael Drayton's *Noahs Floud* (1630) bemoans the permanent loss of "gigantic trees" (Swindell 43). Such ideas are further explored in Olmi's film *Genesi: la creazione e il diluvio* (*Genesis: Creation and Flood*, 1994), where Noah tends an onboard nursery.

Other adaptations explore human efforts to maintain environmental balance. A crucial command from the *Atrahasis Epic* is "destroy your house; build a boat" referring to a house "made of reeds, strong and willowy, that can easily be recycled to plait a lifeboat" (Finkel 118). The *Gilgamesh Epic* reiterates this refrain. Additionally, it includes a cautionary tale of a king coming to regret despoiling the jungle. In *Genesis Rabbah*, God instructs Noah to be "self-sustaining" by planting the trees to use as wood for the ark (Neril and Dee 21), which also appears in Islamic texts such Al-Tha'labī's *Ara'is Al-Majalis Fl Qisas Al-Anbiya* (*Lives of the Prophets*, c. 1000) and the seventeenth-century *Hayat al-Qulub* (*Stories of the Prophets*). Neril and Dee also find a "reference to organic fertilizer" in another passage from *Genesis Rabbah* discussing animal waste (20). Disney's animated short *Noah's Ark* (Justice, 1959) not only recycles the story, but also creates its animals from recycled household materials such as corks, pegs, pencils, and pipe cleaners. In general, then, ecophilic elements of the story have gradually increased as the flood story has evolved over time.

Ecophobia in Flood Narratives

While there has been a general increase in ecophilic aspects of the flood story, the degree of ecophobia has been less consistent. In both *Atrahasis* and *Gilgamesh* the floods are caused by gods seeking to reduce human numbers. In *Atrahasis* this is because humans are too noisy; however, in Gilgamesh the cull seemingly concerns the environmental consequences of overpopulation, given themes of living in harmony with nature and humanity's negative environmental impact. Genesis changes this proto-Malthusian motivation. God acts against "wickedness... evil... violence," not overpopulation (*New Revised Standard Version*, Gen. 6:1–13). Indeed, after the flood God commands Noah and his descendants:

"Be fruitful and multiply, and fill the earth" (Gen. 9:1), repeating his original edict in Genesis' opening chapter and subverting *Gilgamesh*'s and *Atrahasis*' desire to limit the human population. Once the "evil" population has been purged, Genesis' god desires Earth be refilled. Its flood story ends with a list of Noah's descendants demonstrating how this mandate was fulfilled. Moreover, Genesis continues: "The fear and dread of you shall rest on every animal of the earth, and on every bird of the air, on everything that creeps on the ground, and on all the fish of the sea; into your hand they are delivered" (9:1–2). This introduces a fundamental, ecophobic, antagonism, which persists in later Judeo-Christian evolutions of the story such as *The Book of Jubilees* (second century BCE) and the late-mediaeval *Sefer ha-Yashar* (Book of Jasher), but which is absent in Qur'anic versions. Furthermore, Genesis has God proclaiming that humans shall "have dominion... over every living thing" (1:28), words that have been used by some Christians to justify exploiting the natural world, particularly since "Francis Bacon, in the seventeenth century,... hijacked the Genesis text to authorize the project of scientific knowledge and technological exploitation" (Bauckham 6). Giving humans dominion over the nonhuman world is arguably the ultimate act of anthropocentric exceptionalism.

One of the most persistent ecophobic elements pervading these adaptations is god(s) judging, and therefore destroying, the world. A supposedly morally perfect god consigning the world to destruction removes treatment of the nonhuman world from the bounds of morality. This concept has led some Christians to embrace what Fretheim terms a "theology of demolition," essentially the idea that "the world is going to end soon anyway, so there is little need to be concerned about the environment" (683). This sense of doom is echoed in the only place in the Bible where Jesus mentions Noah, comparing "the days of Noah" with a future day of judgment (Luke 17:26–35 and paralleled in Matt. 24:34–42). The apocalyptic tone of the passage is often read as a permanent act of environmental destruction, which many branches of American evangelicalism believe Christians will escape. Such beliefs correlate with attitudes to the nonhuman world. The 2023 American PRRI survey found 62% of White evangelical Protestants consider environmental disasters as evidence of "the end of times," while only 31% agree that "climate change is caused mostly by human activity" (compared to 61% of all Americans) ("The Faith Factor"). The flood story, then, remains vulnerable to ecophobic adaptive innovations.

The Flood "Hero"

The story's central character, whose name changes from Ziusudra to Atrahasis to Utnapishtim before settling on Noah/Noach/Nuh, varies significantly. In the earliest Mesopotamian versions his heroism derives from opposing gods. Christian- and Islamic-inspired adaptations take a more positive view of Noah. A single *New Testament* description of Noah as a "herald of righteousness" (2 Peter 2:5) developed into stories of him preaching to potential flood victims. The first book of the *Sibylline Oracles* (c. 600 CE) devotes a third of its Noah material to his preaching (186–243). Similarly, *The Qur'an* (611 CE) portrays Nuh as a proto-Muhammad figure whose extensive preaching meets with skepticism. This develops in the later *Ara'is Al-Majalis Fl Qisas Al-Anbiya,* where Nuh endures savage beatings for preaching (Brinner 92). Noah's sanctification continues as adaptations progress: Michael Drayton's *Noahs Floud* (1630) "gives us a preachy Noah" (Swindell 43), Milton's *Paradise Lost* (1667–74) describes him as "*perfet*" (875–9); Noah thus becomes an archetype of Christ.

Yet Judeo-Christian portrayals of Noah are not uniformly positive. *Sanhedrin 108a* and *Tanchuma Noach* (500–800 CE) argue he might be less perfect relative to later generations. Some Jewish writers criticized his perceived passivity in Genesis, accepting humanity's fate and obediently building the ark. The "traditional" Jewish view today does not consider Noah righteous, because righteousness is demonstrated by what you do for others, and Genesis' Noah does not help his neighbors. *Tanchuma Noach* also describes Noah missing sleep due to feeding duties (s9). Such "sleep-deprivation offers a rich opportunity to explore Noah as a character on the edge of sanity" (Lilly 33).

The Flood on Film

Having survived 4000 years, inevitably the flood story was soon adapted by the new medium of cinema. The last 125 years have witnessed Noah adaptations from early silent era stop-motion to massive epics, via Egyptian claymation, children's cartoons, and multi-episode Brazilian telenovelas. Following the first live-action Noah film, *The Deluge* (Strauss) in 1911, came a Hollywood blockbuster, *Noah's Ark*, in 1928 (arguably America's first disaster movie) whose torrential flood waters far exceeded those of previous versions. Later adaptations include *The Green Pastures* (Connelley and Keighley, 1936), where Noah haggles with God about how much liquor he can stow onboard the ark and *The Bible: In the Beginning...* (Huston, 1966).

In the late twentieth and early twenty-first centuries, on-screen portrayals of Noah have accelerated rapidly. Half of the fifty-eight productions I have uncovered were produced after 1992 (Page, "'Full' List"). While animated/children's films account for many of these, this century has seen several feature-length, live-action treatments, including two big-budget releases by major Hollywood studios. *Evan Almighty* (Shadyac, 2007) and *Noah* (2014). Both use the flood as a metaphor for climate change, indicative of cultural anxiety around this issue.

Noah (2014)

Aronofsky's film continues in the tradition of many of the elements discussed above and supplements the biblical account with material from later Jewish adaptations, such as the *Book of Enoch* (c. 200 BCE) and *The Zohar* (thirteenth century), along with other representations such as Doré's *The Deluge* (1866). Aronofsky's own prior interpretations of the story also play a part, including his 2011 graphic novel, *Noe: pour la cruaute des hommes*. The film's Noah sees visions of the coming destruction, so he builds an ark. Once his family and a selection of animals are onboard, his sole daughter-in-law gives birth to twin girls. Noah comes to believe "The Creator" (as he calls God throughout) wants him to kill the twins, ensuring humanity's annihilation, but Noah cannot. Once the waters recede he is wracked with guilt, goes on a drinking spree, and struggles to overcome the rift with his family.

Aronofsky's earliest statements about the film were to describe Noah as "the first environmentalist" and the story as an "environmental apocalypse," suggesting the film would adopt an ecophilic perspective (Sciretta). The film contains several ecophilic elements. For Aronofsky and his writer Ari Handel, humanity is being judged because of its failure to care for the world, combining Genesis's judgment for corrupting the earth with *The Gilgamesh Epic*'s concern about Gilgamesh's reckless disregard for the environment. Humanity's negative impact on environmental balance is established from the start and personified in Tubal-Cain, whose followers over-mine the precious substance *tzohar*, gut the forests, and kill animals to eat. In contrast, Noah lives in harmony with the nonhuman world. His (repeated) creed—"The Creator made Adam in his image and then placed the world in his care"—emphasizes a proactive approach to protecting the earth. Thus the film creates a good versus evil dichotomy between the two men, effectively pitting ecophilia and ecophobia against one another. This is reflected in the way the words from Genesis are split and set against each other. Noah emphasizes "be fruitful and multiply and replenish the earth," whereas Tubal-Cain argues that having created the earth, The Creator needed someone "to take dominion over it and subdue it."

Initially, Noah is presented as an exemplar for living sustainably. He forages for food, prevents Tubal-Cain's men from killing animals, and even stops his son from picking a flower, explaining "we only collect what we can use and what we need." Similarly, just as earlier adaptations had Noah sustainably planting the trees he needed to build the ark, here he plants a magical seed, which causes a small forest to grow almost instantaneously, providing the trees he requires. A further ecophilic aspect of the film was Aronofsky's decision to use CGI for the film's animals. Having been "alarmed" at the living conditions for the live primates on his film *The Fountain* (2006), Aronofsky realized it would be "against the actual themes of the film" to start "taking sentient creatures and sticking them on a set" ("Darren Aronofsky Reveals").

Yet despite Aronofsky's apparent good intentions, various elements of the film are ambivalent or contradictory, perhaps even ecophobic, weakening the film's environmentalist credentials. Prior to the flood, Noah explains that "men are going to be punished for what they've done to this world," but this slowly morphs into Noah believing "his divinely given role is to ensure the destruction of the human species" by killing his granddaughters (Reinhartz 290). Ultimately, his "transition from justice to mercy" (to use Aronofsky's words) comes in time to save the twins, providing the audience with some respite (Greydanus). However, the film's underlying logic suggests such mercy is "doomed to fail" (Page, *100 Bible Films* 191). Inevitably, humanity will, once again, threaten the nonhuman world.

Other well-intended elements of the film also fail to follow through. Firstly, given their barren surroundings, it is unclear how Noah's family find sufficient food by foraging alone. Secondly, Noah's construction of the ark leaves his miraculous forest looking almost as barren and ravaged as Tubal-Cain's city. Finally, Samuel Tongue argues that using only CGI animals "confirm(s) an anthropocentric moral outlook," removing the animals' agency (192). The animals' agency is also minimized within the film by Noah anaesthetizing them.[6]

With $359 million in worldwide ticket sales, *Noah* became Aronofsky's highest grossing film ("Noah"), and many religious believers, including evangelicals, appreciated Aronofsky's ecological awareness. However, it was heavily criticized by some religious commentators, particularly American evangelicals. Screenwriter Brian Godawa's description of Noah as an "environmentalist wacko" typified many objections. This disapproval of the way *Noah* portrays both God and Noah as attempting to save the world from humanity's destructive influence seems to correlate with more ecophobic interpretations of Genesis, such as man's dominion over the earth and God's ultimate destruction of the planet.

[6] In contrast, John Huston emphasized animal agency. After experiencing an elephant spontaneously pulling him closer with her trunk, Huston re-created the moment in a shot featuring Noah (317).

Noah (2014) and Evan Almighty (2017)

If clusters of adaptations within a short time frame suggest specific cultural anxieties, what can be observed by comparing *Noah* with its near contemporary *Evan Almighty*? The genre, tone, and seriousness of the two films are wildly different. *Evan Almighty* is a fantasy comedy that reimagines the flood story in a modern context, while foreshadowing a potential future climate catastrophe. Evan Baxter is a newly elected congressman whom God tells to build an ark. Baxter faces widespread mockery but is proved right, becoming a flood hero when a dam bursts, flooding his local area.

As well as shared source material and culture, both adaptations focus on humanity's relationship with the nonhuman world. Animals play an even greater role in *Evan* than in *Noah*, even helping build the ark. Yet both films are also products of capitalism, created for consumption on a massive scale to bring a profit for shareholders. An inherent tension exists between these two ideologies. For example, *Evan Almighty* may make the capitalists the antagonists but, ultimately, when confronted with the evidence, they succumb. Similarly, the film's critique of capitalism surrounds a specific local dam rather than ongoing operations consistently releasing harmful gases into the planet's atmosphere.

Further analysis of these films highlights these tensions. In his discussion of ecological disaster movies, Robert Geal draws attention to the Cartesian nature of most Hollywood cinema, "prioritising humanity and devaluing nature," through a cinematic language that "continuously reposition[s] the spectator as master of meaning and space" (*Ecological* 15, 49). Films like *Noah* juxtapose "narratives warning about ecological disaster" with an "anthropocentric Cartesian grammar" (77). Such movies "repress anxiety about ecological degradation" (72) by presenting themselves as pleasurable entertainment and allow the lead characters to survive the disaster. By soothing/repressing this anxiety, the urgency to take real-world action to combat climate change is potentially dissipated.

These tensions are also reflected in Noah's and Evan's character arcs. Noah is an environmental extremist who stops his child picking flowers and entertains the possibility of killing his grandchildren to prevent future environmental destruction. Yet ultimately, he relents, "learns about mercy as well as justice" (Greydanus), is reintegrated back into his nuclear family, and then drinks to excess (consuming more than he *needs*). Essentially, the activist is tamed/compromised. In contrast, Evan starts as a casually ecophobic, but relatable, everyman who learns about the importance of protecting the nonhuman world. Having started from opposing positions, both characters end up in a middle ground between extreme lifestyles and irresponsible ones. So if the formal elements of these films reassure audiences that they too can survive, they also suggest that doing so will not require us to change too radically, essentially suggesting green capitalism, rather than a more restricted lifestyle, might be sufficient to avoid a climate catastrophe, despite evidence to the contrary.

Staying Afloat

The title of this chapter echoes the line famously attributed to Martin Luther King that "the arc of the moral universe is long, but it bends toward justice," with the hope that it might also deviate toward environmental justice, delivering a more equitable and sustainable relationship with the nonhuman world. Yet when we see how this story has been reused and recycled over 4000 years, the evidence is less optimistic. At times the myth of the great flood has been adapted to express a greater affinity with the nonhuman world, but at others it has been deployed in a more ecophobic manner.

At the time of writing, Aronofsky's *Noah* is a decade old. Since its release new adaptations of the flood story have emerged, such as Robert Savo's *Noah and the Flood* (2021), which draws on visual ideas from Aronofsky's film, and *Noah's Ark: A Musical Adventure* (2024), which maintains the link between the ancient flood and climate change in our own era. What we find is flood adaptations repeatedly resurfacing to express greater affinity with the nonhuman world than was typical in the past. This highlights the cultural anxiety around climate change. Even when a series of adaptations swerve toward ecophobia, nonhuman elements run deep enough to eventually re-emerge. The story is one that readily evolves to allow those adapting it the opportunity to respond to significant cultural anxiety in times of unprecedented threats—be they geographical or metaphorical—even if such stories struggle to liberate themselves from anthropocentrism. While these films may soothe, or perhaps suppress, these anxieties, it is unclear whether they are part of the solution to repairing the damage to the nonhuman world, or a release valve holding us back from making the necessary changes.

In 2008 a new version of the Mesopotamian flood story emerged, revealing that Atrahasis' original craft was a circular coracle. "A circular coracle is exceptionally buoyant," Finkel observes, "and if it happened to be difficult to steer... that would not matter, because all it had to do was keep its precious contents safe" (125). Much as we might desire to see the story of Noah's Ark set sail for ecophilic shores, its future direction is uncertain. Somehow, like Atrahasis' coracle, it is built just to survive.

4000 Years of Adaptations (Dates are Approximate)

Sumerian Deluge (2000 BCE)
Atrahasis Epic (1750 BCE)
Gilgamesh Epic (650 BCE)
Genesis (600 BCE)
Book of Enoch (200 BCE)
Book of Jubilees (200 BCE)
Genesis Apocryphon [1Q20/1QapGen] (100 BCE)
Pesher/Commentary on Genesis [4Q252/4QPGen] (50 BCE)
Flood Apocryphon/Admonition Based on the Flood [4Q370] (40 BCE)
Pseudo-Philo (Liber Antiquitatum Biblicarum) (80 CE)
Josephus, *Antiquities of the Jews* (94 CE)
Bereshit of Rabbah (Genesis Rabbah) (400 CE)
Sanhedrin 108b (Babylonian Talmud) (500 CE)
Sibylline Oracles (600 CE)
The Qur'an (610 CE)
Midrash Tanchuma Noach (750 CE)
Abu Ishaq al-Tha'labi. *Ara'is Al-Majalis Fl Qisas Al-Anbiya* (Lives of the Prophets) (1000 CE)
Sefer ha-Yashar (Book of Jasher) (1000 CE)

The Zohar (1270 CE)
Unknown. *Noye's Fludde* (Chester Mystery play) [Play] (1467)
Hieronymus Bosch. *Noah's Ark on Mount Ararat* [Painting] (1510)
Jacopo Bassano. *The Animals Entering Noah's Ark* [Painting] (1570)
Jan Brueghel the Elder. *The Entry of the Animals into Noah's Ark* [Painting] (1613)
Michael Drayton. *Noahs Floud* [Poem] (1630)
Giovanni Benedetto Castiglione. *In Front of Noah's Ark* [Painting] (1650)
Andrew Marvell. *Upon Appleton House* [Poem] (1651)
Van den Vondel, Joost. *Noah or the Downfall of the First World* [Play] (1667)
John Milton. *Paradise Lost* [Poem] (1674)
John Dryden. *Absalom and Achitophel* [Poem] (1681)
Allamah Muhammad Baqir Majlisi. *Hayat al-Qulub* (*Stories of the Prophets*) [Hagiography] (1699)
Francis Danby. *The Deluge* [Painting] (1840)
Gustave Doré. *The Deluge* [Woodcut] (1866)
Arthur Melbourne Cooper. *The Tale of the Ark* [Film] (1909)
Richard Strauss. *The Deluge* [Film] (1911)
Michael Curtiz. *Noah's Ark* [Film] (1928)
Marc Connelly, William Keighley. *The Green Pastures.* [Film] (1936)
Benjamin Britten. *Noye's Fludde* [Opera] (1958)
Bill Justice. *Noah's Ark* [Film] (1959)
John Huston. *The Bible: In the Beginning...* [Film] (1966)
Julian Barnes. *A History of the World in 10½ Chapters* [Novel] (1987)
Ermanno Olmi. *Genesi: la creazione e il diluvio* (*Genesis: Creation and Flood*) [Film] (1994)
Tom Shadyac. *Evan Almighty* [Film] (2007)
Juan Pablo Buscarini. *El Arca* [Film] (2007)
Darren Aronofsky, Niko Henrichon. *Noe: pour la cruaute des homme* [Graphic novel] (2011)
Darren Aronofsky. *Noah* [Film] (2014)
Robert Savo. *Noah and the Flood* [Film] (2021)
Alois Di Leo, Sergio Machado. *Noah's Ark: A Musical Adventure* [Film] (2024)

Works Cited

Bailey, Lloyd R. *Noah: The Person and the Story in History and Tradition*. U of South Carolina P, 1989.

Bauckham, Richard. *Bible and Ecology: Rediscovering the Community of Creation*. Baylor UP, 2010.

Brinner, William M. *Ara'is al-Majalis fi Qisas al-Anbiya' or "Lives of the Prophets" as Recounted by Abu Ishaq Ahmad ibn Muhammad ibn Ibrahim al-Tha'labi*. E. J. Brill, 2002.

Carr, David McLain. *Reading the Fractures of Genesis: Historical and Literary Approaches*. Westminster John Knox P, 1996.

Collins, Matthew A. "An Ongoing Tradition: Aronofsky's *Noah* as 21st-Century Rewritten Scripture." *Noah as Antihero: Darren Aronofsky's Cinematic Deluge*, edited by Rhonda Burnette-Bletsch and Jon Morgan, Routledge, 2017, pp. 8–33.

"Darren Aronofsky Reveals the Secret to Noah's Realism in Exclusive PETA Video." *PETA*, 25 Mar. 2014. www.peta.org/media/news-releases/darren-aronofsky-reveals-secret-noahs-realism-exclusive-peta-video/. Accessed 10 Aug. 2025.

Finkel, Irving. *The Ark Before Noah: Decoding the Story of the Flood*. Hodder, 2014.

Fretheim, Terence E. "Genesis and Ecology." *The Book of Genesis: Composition, Reception and Interpretation*, edited by Craig A. Evans, Joel N. Lohr, and David L. Petersen, Brill, 2012, pp. 683–706.

Gaster, Theodor Herzl. *Myth, Legend, and Custom in the Old Testament: A Comparative Study with Chapters from Sir James G. Frazer's Folklore in the Old Testament*. Harper & Row, 1969.

Geal, Robert. *Ecological Film Theory and Psychoanalysis*. EPub. Routledge. 2021.

Geal, Robert. "Towards an Ecocritical Adaptation Studies." *Adaptation*, vol. 16, no. 1, Mar. 2023, pp. 1–12. https://doi.org/10.1093/adaptation/apad001.

Godawa, Brian. "Darren Aronofsky's Noah: Environmentalist Wacko." *Godawa's Movie Blog: Hollywood Worldviews*, 29 Oct. 2024. https://web.archive.org/web/20250323171802/godawa.com/darren-aronofskys-noah-environmentalist-wacko/. Accessed 30 Nov. 2025.

Greydanus, Steven D. "Interview: 'Noah' Writer-Director Darren Aronofsky and Co-Writer Ari Handel." *National Catholic Register*, 21 Mar.

2024. www.ncregister.com/news/interview-noah-writer-director-darren-aronofsky-and-co-writer-ari-handel. Accessed 10 Aug. 2025.

Huston, John. *An Open Book*. Macmillan, 1981.

Leitch, Thomas. "History as Adaptation." *The Politics of Adaptation: Media Convergence and Ideology*, edited by D. Hassler-Forest and P. Nicklas, Palgrave Macmillan, 2015, pp. 7–20. https://doi.org/10.1057/9781137443854_2.

Lilly, Ingrid E. "Rock Giants and the Magic Stone of Torah: Allusions to Esoteris and Extra-Biblical Literature in Noah." *Noah as Antihero: Darren Aronofsky's Cinematic Deluge*, edited by Rhonda Burnette-Bletsch and Jon Morgan, Routledge, 2017, pp. 34–48.

King, Martin Luther. "Our God is Marching On!" 25 March 1965, Montgomery, Alabama. Speech.

Midrash Rabbah. Translated and edited by H. Freedman and Maurice Simon, Soncino P, 1961.

Milton, John. *Paradise Lost*. Collier-MacMillan, 1962.

Mottram, Stewart. "Deluge and Disease: Plague, the Poetry of Flooding, and the History of Health Inequalities in Andrew Marvell's *Hull*." *The Seventeenth Century*, vol. 38, no. 2, 2023, pp. 263–290. www.tandfonline.com/doi/full/10.1080/0268117X.2022.2142656. Accessed 10 Aug. 2025.

Neril, Yonatan, and Leo Dee. *Eco Bible: Volume 1: An Ecological Commentary on Genesis and Exodus*. The Interfaith Centre for Sustainable Development, 2020.

"Noah." *Box Office Mojo*, undated. www.boxofficemojo.com/releasegroup/gr3025359365/. Accessed 10 Aug. 2025.

Noah's Ark. Directed by Michael Curtiz, story by Darryl F. Zanuck, Warner Brothers, 1928.

Page, Matthew. *100 Bible Films*. BFI/Bloomsbury, 2022.

Page, Matthew. "'Full' list of Noah and the Ark films." *Bible Films Blog*, 21 Sep. 2024, biblefilms.blogspot.com/2024/09/full-list-of-noah-and-ark-films.html. Accessed 10 Aug. 2025.

Puchner, Martin. *Literature for a Changing Planet*. Princeton UP, 2022.

Reinhartz, Adele. *Bible and Cinema: An Introduction*. Routledge, 2013.

Rhys, Ernest. *"Everyman," with Other Interludes, Including Eight Miracle Plays*. J. M. Dent, 1909.

Sciretta, Peter. "Interview: Darren Aronofsky—Part 3." */film*, 12 Sept. 2008. web.archive.org/web/20081002192030/http://www.slashfilm.com/2008/09/12/interview-darren-aronofsky-part-3/. Accessed 10 Aug. 2025.

Shepherd, David. "'Noah's Beasts Were the Stars': Arthur Melbourne Cooper's *Noah's Ark* (1909)." *Journal of Religion & Film*, Vol. 2, Issue 1, Article 20, Jan. 2016. https://digitalcommons.unomaha.edu/jrf/vol20/iss1/20. Accessed 10 Aug. 2025.

Stavrakopoulou, Francesca. *God: An Anatomy*. Picador, 2021.

Swindell, Anthony C. *Reworking the Bible: The Literary Reception-History of Fourteen Biblical Stories*. Sheffield Phoenix P, 2010.

The Bible. New Revised Standard Version, Thomas Nelson, 1989.

"The Faith Factor in Climate Change: How Religion Impacts American Attitudes on Climate and Environmental Policy: Findings from the 2023 Climate Change Survey." *PRRI,* Washington, D.C., 2023.

Tongue, Samuel. "'It's Not the End of the World': Aronofsky's *Noah* and ImaXed Apocalyptic Animals." *Noah as Antihero: Darren Aronofsky's Cinematic Deluge*, edited by Rhonda Burnette-Bletsch and Jon Morgan, Routledge, 2017, pp. 183–98.

8

"Posters of the Sea and Land": Indexing Environmental Change in Alexander Abela's Ecological Adaptation of *Macbeth*

Allen H. Redmon

By 2021, international agencies were declaring an unprecedented emergency in southern Madagascar. The United Nation's World Food Programme (WFP), for instance, warned that back-to-back droughts in the southern part of the island were causing the first famine in "modern history to be solely caused by global warming" (Baker). The lack of rain in the region had pushed already higher natural air temperatures even higher as carbon pollution intensified temperatures beyond their natural levels. Historically dry climates became even drier. The wind became more constant. These two effects led to less moisture in the soil, which made it even more difficult to grow nutrition-dense foods. Worse, soil was turning to sand, and sand was washing over or at least displacing entire villages. Communities that might otherwise be able to turn to the sea were being driven to other parts of the island, too, as warmer waters were pushing fish swim patterns even further from the coastline. Warmer waters around the globe were also leading to sea rise, which

A. H. Redmon (✉)
Texas A&M University Central Texas, Killeen, TX, USA
e-mail: allen.redmon@tamuct.edu

P. Demory (ed.), *Ecoadaptation*, Palgrave Studies in Adaptation and Visual Culture,
https://doi.org/10.1007/978-3-032-09847-4_8

meant that otherwise habitable lands were being submerged beneath the sea. The WFP led efforts like "dune-fixing" to try to slow the loss, but the forfeiture continued at alarming rates (Rahmoun).

The irony, of course, as it relates to the climate crisis behind these changes is that Madagascar contributes very little to the emissions fueling this local situation. The independent research group Climate Action Tracker finds that Madagascar produces one of the lowest levels of carbon dioxide equivalent (MtCO2e) in the world. Scientists estimate Madagascar emits only 0.1 metric tons of MtCO2e per person. By comparison, the United States emits 19.0 metric tons of MtCO2e per person, Australia emits 20.2 metric tons, and Saudi Arabia emits 20.7 metric tons. Collectively, China, the United States, India, and the European Union emit "83% of all total emissions." The data show that developed countries have "built their economies—and their wealth—on" practices that increase gashouse emissions across the globe, and the effects of these emissions impact far-off places that produce very low emissions and lack wealth—like Madagascar (Paddison and Choi).

The current climate situation in Madagascar changes how one watches Alexander Abela's *Makibefo* (1999/2001). Rather than simply an auteur-driven instance of global Shakespeare (see Burnett) or a thoughtful transcultural adaptation (see Bomnüter), set within the climate crisis, *Makibefo* functions as an indexical marker of the changes that have occurred since the filming of this project in the southern part of Madagascar. Abela's film develops into a visual version of what the three witches declare themselves to be in Shakespeare's *Macbeth*, "posters of sea and land," which is to say couriers of some news that needs to be delivered with some sense of urgency (1.3.34). Abela layers onto Shakespeare's story of unchecked ambition an exposé on the environmental implications a pursuit of power causes even those parts of the world unentangled in the behavior that pushes the world toward an existential crisis. Abela adds several visual and narrative flourishes that suggest the director was aware of this reading even as he was making his film more than twenty-five years ago, which is to say slightly ahead of the current crisis. This chapter highlights these elements to consider the way in which this underappreciated film can serve as an ecological marker of climate change in southern Madagascar.

Shakespeare and Climate Change

One might be surprised to learn that this reading of Abela's film actually fits an attention to weather found across Shakespeare's plays. This chapter begins with a discussion of this feature of the Bard's work that uses Sophie Chiari's recent monograph on references to weather and climate in Shakespeare. Her work suggests strategies for an ecological reading of an adaptation of *Macbeth*, even if the play is not often discussed in this manner. Chiari demonstrates that people in the Elizabethan and Jacobean periods were dealing with shifting attitudes about the weather and climate that match our own debates. Attention to these equivalencies can help to depoliticize the discussion around climate, enabling a new way to understand the ecological realities developing around the globe. Viewing *Makibefo* through the lens of Chiari's work focuses our attention on the setting and soundtrack in Abela's film, which is regularly filled with the sights and sounds of sand, wind, and waves. Set within Chiari's insightful analysis, these images and sounds become as important to the adaptation as the narrative choices Abela makes.

Ecocritical studies of Shakespeare's plays have not gone far enough in their analysis of weather and climate, according to Chiari. Most studies note the extent to which references to weather or climate serve as metaphors or analogies for serious discussion on a range of societal ills, but they do not establish the ways in which these references can be "an end in [themselves]" (18). Chiari supports her argument that these references can be taken literally by detailing the regular attention scientists, theologians, farmers, politicians, and, really, people from all walks of life in England gave weather and climate during the Reformation, Elizabethan, and Jacobean periods. Chiari identifies several reasons for this broad and sustained interest. For one thing, humans were newly grappling with their relationship to the weather, which had long been thought to be up to the whims of the gods. Now, "men of letters" began to propose ways that people could "act upon celestial forces rather than being acted upon by them" (1). Such proposals would have appealed to those in England during Shakespeare's time since England had been enduring particularly harsh conditions and "several atmospheric shifts" (3). In what is now referred to as the "Little Ice Age," "the second half of the sixteenth century and the first half of the seventeenth century proved especially harsh periods" during both summer and winter (Chiari 3–4). There were, of course, those who attributed these conditions to the gods. But others saw something more natural. The debate itself, Chiari reasons, encouraged those who entered it to develop a more precise meaning of terms like environment, weather, and climate.

The term "environment" during the early modern period would have already been closely associated with the concept of nature, according to Chiari, which people during that period would have been more prone to see as something that impacts them and that they could impact. Chiari suggests that this was a significant shift in perception: "nature was no longer exclusively perceived as God's creation and... Europeans engaged with it in fresh ways" (7). The weather, taken to be fits of heat or cold, wind and rain, or, in good times, temperate conditions, began to reflect more than the attitudes or dispositions of the gods. They became natural occurrences that could be stretched across time and fit within perceivable patterns. Scientists began to distinguish one climate from another and to ascribe some routine to what might otherwise have been deemed purely random and arbitrary. The climate became, then, something of an organizing principle rather than a backdrop.

Chiari illustrates her claim by tracing the way references to weather provide some structure to every type of Shakespearean play. Comedies like *A Midsummer Night's Dream* and *As You Like It* rely, respectively, on concepts of "wetness" and "cold" (19–20). In the case of the former, Chiari contends that references in *A Midsummer Night's Dream* to humidity, rain, and mud, and to the fact that these conditions often follow human actions, "challenge medieval beliefs [about the weather] and... test fresh hypotheses, such as the idea that people's attitudes may in fact be responsible for climatic imbalance" (20). References to "cold" in *As You Like It* work similarly, "suggest[ing] frozen and sterile thoughts," which undercut the celebrations a more surface reading of references to weather would emphasize (20). Tragedies also employ weather in unexpected ways. The words "hate" and "heat," for instance, are interrelated through *Romeo and Juliet* in a way that "overdetermines the climate of the play" (20). And *Othello* is "obsessed with breath and wind," which turns the play into a "cosmological piece" reflecting on the ways "climate and air" can "make [one] the victim of [one's] own humours" (20–21). The opposite occurs in *King Lear*, as Lear himself is depicted as a "raging storm striking the heath" (21). In *Anthony and Cleopatra* Chiari suggests something of a balance between the two positions sketched in *Othello* and *King Lear*. The titular characters become "explorers of sky and heavens," with "tears, jealousy, cruelty and burning pain [being] conveyed through imagery of gales, hail, and tempests" (21–22).

Such references, Chiari reasons, revised and reapproached at various times and developed across a career, culminating in *The Tempest*, reveals that Shakespeare had not some passing or poetic interest in the weather, but rather an extended consideration of "the serious incidence of environmental pitfalls on human lives and, conversely,... the disasters resulting from men and women's interference with their natural environment" (22). Chiari's final position brings Shakespeare's wrestling with the weather and climate into present-day considerations. The playwright is already staging the contemporary debates on climate change. It makes sense, then, that audiences would find an ecological statement in an adaptation of a Shakespearean play. Alexander Abela's *Makibefo* turns out to be just this sort of statement.

Sea, Sand, and Sky in *Makibefo*

At first blush, viewers are likely to simply delight themselves in the equivalencies Abela's plot has with Shakespeare's and the way these equivalencies find new meaning when dressed as they are in Antandroy culture and practice. The film's plot approximates key points in *Macbeth*, and Abela makes the parallels explicit by having a storyteller (Gilbert Laumord) admit and roughly summarize the events the film will stage. He begins the film with the words, "In a land washed by the ocean, a tribe of people lived in sight of sands and crashing waves. Their king was a noble king, who gave his people peace and harmony. And amongst his subjects many were good and true. But none more so than Makibefo." The storyteller continues to relate the story that will unfold when the plot begins most properly: Makibefo (Martin Zia) captures a fugitive.

On his return to his village, Makibefo encounters a witch doctor (Victor Raobelina), who signifies that Makibefo will be king. Makibefo and his wife, Valy Makibefo (Neoliny Dety), conspire to kill the tribe's leader. They do so, framing his guards. Makibefo later has Bakoua (Randina Arthur) killed and still later kills the family of Makidofy (Jean-Noël), all of which follows key points in *Macbeth*. So, too, do the results of these ill-fated actions. Valy Makibefo goes mad and drowns herself in the sea. Makibefo waits on the shore for Makidofy and his companions to kill him. It is, like *Macbeth*, and as the storyteller says at the end of his opening monologue, "a story of damnation."

Given the location of the film, the prominence of the sounds of the sea, and the movie's visual imagery, the story of damnation Abela relates extends beyond the plot to the impact of climate change on places like the southern tip of Madagascar near Faux Cap where *Makibefo* was filmed. Abela signals from the opening moments of his film that he means his project to highlight natural forces. His movie stages something of a competition between land and sea that ultimately focuses on the places where these two topographies meet. The water is always subsuming those who get near it, meaning that the competition between land and sea is always tilted toward the sea. This is true from the opening moments of the film, which are heard rather than seen. Abela's soundtrack delivers the sound of rushing waves even before the image of a beach emerges from the blackness that initially covers the screen. When the opening image is fully revealed, Abela still favors the oceanic waters by having the images of the sea and sky cover roughly seventy-five percent of the screen. The beach is quite literally pushed in this opening image to the left edge of the screen. Abela intensifies this imbalance, too, when he sets along the beach four *aloalo* poles—tall, thin ancestral sculptures used by the Mahafaly tribe in Southern Madagascar during funerals to mark the increase of the generations over time. Abela's placement of the poles set against the contest of land and sea loads these structures with at least two unexpected points of significance (Fig. 8.1).

First, in terms of the story Abela adapts, the four poles suggest *Macbeth's* overt attention to the tension between timelessness and transience. The pole set furthest to the right on the screen extends the horizontal length of the screen. Abela tempts his viewer to imagine the

Fig. 8.1 Abela opens the film with an image of four *aloalo* poles across a beach that is contending with the sea and sky for space on the screen

sculpture stretching both to the very depths of the earth and the very top of the heavens. Upon reflection, one might associate this pole with the desire the drama's leading couple has for a never-ending legacy. The other three poles stand to the left of the most prominent pole. The middle pole would seem to be grounded in the bottom of the frame. It extends toward the top of the screen before being capped by an open-ended piece that breaks the pattern of an otherwise self-enclosed decorative pole. The two poles that flank this center pole are similarly capped. Visually, they differ from the pole at the center of their grouping by having a clear beginning and ending. Together, in keeping with a thematic interpretation of the poles, the three poles might stand as the three weird sisters, whose words open Shakespeare's play and remind the audience as the play unfolds of the certainty of some final stop to all, even those who have been destined for a time to rule.

Secondly, the opening images of the sea, sky, land, and *aloalo*-marked beach evoke a more general significance, one that this chapter argues might very well establish the most unmistakable connotation of Abela's film twenty-five years after its release. In this current moment, one where the very sands Abela brings to the screen have been lost to the sea, the furthest right *aloalo* suddenly stands as a failed barrier between the land and sea. The pole can neither stop the land from slipping into the ocean nor the waters from pushing their way across the vulnerable landscape. The movement of the waters pushing their way right to left across Abela's screen, set against the pronounced sounds of the water in the soundtrack, suggests that the beach is destined to become a sea floor. In this context, the four *aloalo* poles are, in fact, serving their intended purpose as sculptures at a funeral. This service, it so happens, is for the impending loss of a land and a people and all that goes with it.

Abela's attention to climate extends across the entirety of his film. He features the sand, sea, and sky in such a way that these elements very nearly operate as characters in his film, leaning into a contemporary statement on climate change. This happens from the first images following the opening monologue that declares what follows a story of damnation. The first shot within the drama of the film is, initially, a complete white-out before one begins to see an almost indistinguishable mix of sand and sky. A man, later revealed to be a fugitive, Kidoure (Boniface), emerges above the crest of a dune. A tracking shot moves right to left as Kidoure crosses the sand, while another man, Bakoua, gains on him, spear in hand. This chase runs for almost two minutes. As it plays, one hears waves and eventually the sound of deep breaths. Eventually, the camera cuts to Makibefo, who stands looking to the sea before turning his attention to Kidoure, who is always running toward him. Kidoure runs up the dune Makibefo sits atop until he runs into Makibefo's spear, which injures him.

What stands out most during this sequence is the vastness of sand and the sound of the waves and breathing that might well be regarded as the "heavens' breath" mentioned in *Macbeth* (1.6.5). This breath emerges from a man, though, a man who causes the ruin of a tribal order that is otherwise intact at the film's opening even as a fugitive is on the run. The fugitive will be returned. The man will prove uncontrollable. The next sequence extends the threat in the film beyond just one man. Makibefo, Bakoua, and Kidoure cross the dunes to return to their tribal settlement. Their trek has them always walking toward the sea. As they walk, the dunes begin to change. The first barren sands now have vegetation and skeletal remains. The two together reveal the careful balance of life and death at the edge of the sea. This balance can be undone. It will be, and Abela signals that eventual imbalance from the first moments the sea is visible. Abela cuts to an image that he will repeat at key moments in the film. The sea crashes against a large rock some distance from the shore. The sea itself is marked as a threat to the people presently on dry land.

This threat continues to lurk in the background of the scene where the witch doctor delivers his prophecy to Makibefo. The camera keeps the sea in the background the whole time the witch doctor delivers his words. The prophecy itself extends from the scene, at least visually. Abela locates every death in the film after Kidoure's murder, the film's first murder, on the beach and beside the waters. Just as importantly, these waters begin to be associated with Western power. Just after Malikomy kills Kidoure, Abela creates a shot that shows Makibefo running along the beach toward Valy Makibefo. Valy Makibefo is positioned with her back toward the camera on the far left side of the frame (Fig. 8.2).

The camera is unable to capture Valy Makibefo's entire body frame; instead, one sees from the top of her shoulders to the middle of her legs. The framing features her decorative *lamba*, covered in what Bomnüter believes to represent Western power, or "more specifically, *French* royalty: a (French) crown, a (French) Renaissance castle, and the Fluer-de-Lis." In this first shot, Abela reveals most clearly the castle, which fills the bottom left side of the frame. The rest of the frame is filled by water, sand, and Makibefo, who repeats Kidoure's run across the sand in the opening shot

Fig. 8.2 Valy Makibefo (Neoliny Dety) waits dressed in a decorative *lamba* marked with images associated with French royalty as her husband crosses the beach to reach her

by also moving right to left across the screen and, eventually, into his captor. The captor this time, though, is more than a person. She is a person bewitched by Western ideals.

These early shots on the beach mark the beach itself as a place of death. Abela brings every character who dies to the beach before their demise. For instance, Danikany's travel to Makibefo's dwellings is caught in an extreme longshot that shows him and his entourage following the beach at the edge of the ocean. The first image of this procession foregrounds a pirogue sitting in the water that fills the bottom half of the screen. The parade of people moves across the center of the screen with the sand and sky visually above them. The group is quite literally set at the perilous intersection of sand and sea. The sounds of their procession's singing compete with the sounds of the crashing waves as the group marches toward Makibefo and Valy Makibefo. Abela uses the beach similarly just before Danikany's murder. Makibefo sits on the beach at night. The waves enter the screen and then recede as Valy Makibefo drops a knife in front of Makibefo. Once the murder is complete, the couple reunites on the beach. The sound of the waves fills the soundtrack throughout the ordeal. They also fill the soundtrack the next morning when Danikany's murder is discovered. This setting of this action is as important as anything else in the sequence of the murder. The beach becomes not only a point of contact between land and sea, but a site of murder as well.

Abela returns the action to the same setting when Makibefo has Bakoua murdered. The soon-to-be-executed character walks from right to left across the screen along the sea. Abela cuts to the shot of the sea crashing over the lone rock that first appeared when he, Makibefo, and Kidoure arrived at the beach at the beginning of the film. This cut to the rock does more than locate Bakoua alongside the sea. It foreshadows his death and the cause of it. The Western powers of ambition and treachery that mark the lands on the other side of the sea are implicating those otherwise removed from them. Abela highlights this connection by having Bakoua literally walk in the footsteps left from Danikany's recent procession along the same route. Bakoua is eventually flanked on two sides and killed. What is most interesting in the image following this murder is the way the two murderers stand over Bakoua's body. The murderers are multiplying as the action unfolds. One murderer becomes two. Two will become ten by the end of the film.

Before reaching the dramatic ending, Abela stages Valy Makibefo's suicide in what is one of the more interesting aspects of the film as it is being cast in this discussion: she walks into the sea. All that is left of this character after her suicide is her *lamba*, which Makibefo finds in a pool of water. Makibefo holds the garment close to his chest. He squats on the beach in a shot that erases any gap between the sea and the land. There is only seabed. The storyteller reads one of the more well-known monologues from Shakespeare's script as Abela holds the image of the grieving king:

> Tomorrow, and tomorrow, and tomorrow
> Creeps in this petty pace from day to day,
> To the last syllable of recorded time;
> And all our yesterdays have lighted fools
> The way to dusty death. Out, out, brief candle.
> Life's but a walking shadow, a poor player
> That struts and frets his hour upon the stage.
> And then is heard no more. It is a tale
> Told by an idiot, full of sound and fury,
> Signifying nothing. (*Macbeth* 5.5.19–28).

The words carry a wider significance when set in the context of the present discussion. The sound and fury extend beyond a husband and wife, implicating global powers that ignore the impact they are having on the world.

Abela emphasizes this last insight in dramatic fashion during the final moments of the film. The camera cuts to a group of pirogues sailing toward the beach to confront Makibefo. Abela's framing highlights the wood on the boat, which associates these boats with *Macbeth*'s Birnam Wood. Abela does not have the witch doctor include Birnam Wood in his second prophecy. This image does, however, bring that plot point into Abela's plot, albeit in a different light. Rather than a military tactic, this Birnam Wood arises as a metaphor for things like Western expansion and colonization. The resulting image mimics the results of such efforts. Ten men stand alongside Malikomy in front of their pirogues, their backs

Fig. 8.3 Abela closes the film with the recurring image of the rock sitting in the middle of the sea; only now the rock is entirely beneath the waves that only earlier encircled it

to the sea, where the sun is starting to set. They will defeat Makibefo, but their time has been marked. Abela cuts from their triumph to the recurring image of the sea crashing against the rock. The water very nearly washes over the rock entirely. There is only the sea as the film ends (Fig. 8.3).

A Document of a People

Abela's film is not only "a story of damnation" in relation to climate change; it is also a document of an endangered population. Originally, Abela conceived of the project as a documentary but shifted his intentions after he was unable to secure financing. As Anne-Marie Constantini-Cornède recounts, Abela traveled to Madagascar with only one other person on his crew, sound designer Jeppe Jungersen, to make a dramatic film using local people from the Antandroy tribe around Faux Cap as his actors and cinematic collaborators. Abela wanted the film to stand, as Constantini-Cornède describes it, as "a socio-artistic experience," which is to say he wanted the film to stand as both a story and a document of an otherwise unacknowledged part of human existence.

Abela makes this intention explicit when he adds the following intertitle at the end of the film:

> The Antandroy people of Madagascar who played the characters and helped in the making of this film are an ancient tribe with a true sense of pride, honour, and tradition. A poor people in what is already a poor country, they have few possessions and little knowledge of the outside world. As simple fishermen, they live off the ocean that crashes against their unchanging shoreline and take one day at a time. The majority of actors have never seen a television, let alone a film, and have never acted before in their lives.

Abela's film would circulate as an acknowledgment of a specific community living on the edges or even outside of Western society even without this intertitle, but its inclusion draws that quality to the surface for anyone who watches the film again.

One begins to see something other than simply an underfunded project or the work of an auteur interested in establishing himself as "a lone pioneer in the industry, an artist who is remarkable or working independently and against the grain" (Burnett 23). Abela's independence ultimately showcases a people "isolated from the rest of the world" rather than the particularities of an independent filmmaker. As Udo Bomnüter has demonstrated, *Makibefo* serves as "a genuine case of transcultural adaptation... an Antandroy tale told with authenticity and richness in detail, which would never have been achieved without the active participation of the indigenous community." For Bomnüter, Abela's plot extends equally well from Shakespeare's play and Antandroy culture. More impressive than the obvious parallel to *Macbeth* is the clear presence of Antandroy people whose costumes and influence are seen and felt throughout the film. Abela involves the local people as "actors..., (co-) screenwriters, musicians, and crew members." The result is a film "heavily informed by Antandroy tradition." Abela has thus adapted not one but two source texts into one powerful drama that not only explores the demise of a person but locates that demise in an otherwise unnatural ambition that one could argue causes the very climate threat the world currently faces.

Conclusion

Abela's film ultimately justifies Sven Rank's claim that adaptations of *Macbeth*, Shakespeare's plays, and canonical works more generally can open "new *discursive* space for negotiating the future from the textual residues of the past" (14). *Makibefo* subtly asks audiences to grapple with the causes and impact of global climate change and how those impacts are affecting even remote parts of the world. The film uses the story of a singular ill-fated flaw located in a particular husband and wife, what Rank refers to as the "Macbeth kernel," to tell a story of global crisis that adds new cultural conditions to that kernel to give it new life (357). The realization not only supports Rank's ideas that adaptations are contingent on social tensions that demand "a dialectics of strategic or tactical negotiation between cultural locations" but also shows the ways in which cinematic adaptations can lean into representations of broader ecological adaptation. The two together can serve as "posters of the sea and land" that can help us consider again the very real conditions of our current cultural and environmental moment.

Works Cited

Baker, Aryn. "Climate, not Conflict: Madagascar's Famine is the First in Modern History to be Solely Caused by Global Warming," *Time Magazine*, July 20, 2021, time.com/6081919/famine-climate-change-madagascar/. Accessed 29 January 2025.

Bomnüter, Udo. "Tribal Rituals on Sand Dunes: Alexander Abela's *Makibefo* as a Transcultural Appropriation of Shakespeare's *Macbeth*." *Literature/Film Quarterly*, vol. 48, no. 2, Spring 2020. lfq.salisbury.edu/_issues/48_2/tribal_rituals_on_sand_dunes_abelas_makibefo_as_a_transcultural_appropriation_of_shaespeare.html#gsc.tab=0. Accessed 29 July 2025.

Burnett, Mark Thornton. "Alexander Abela," in *Shakespeare and World Cinema*. Cambridge UP, 2013, 23–54.

Chiari, Sophie. *Shakespeare's Representaton of Weather, Climate, and Environment: The Early Modern 'Fated Sky'*. Edinburgh UP, 2019.

Costantini-Cornède, Anne-Marie. "African Kings: Makibefo (1999) and Souli (2004), Alexander Abela's Transcultural and Experimental Screen Shakespeare," in *Shakespeare en devenir*, December 20, 2017, shakespeare.edel.univ-poitiers.fr/index.php?id=1126. Accessed 29 January 2025.

Makibefo. Directed by Alexander Abela, performances by Martin Zia, Neoliny Dety, and Gilbert Laumord, Blue Eye Films, 2001.

Paddison, Laura, and Choi, Annette. "As Climate Chaos Accelerates, which Countries are Polluting the Most?," *CNN*, January 2, 2024. www.cnn.com/interactive/2023/12/us/countries-climate-change-emissions-cop28/. Accessed 29 January 2025.

Rank, Sven. *Twentieth-Century Adaptations of* Macbeth*: Writing between Influence, Intervention, and Cultural Transfer*, Peter Lang, 2010.

Rahmoun, Alice. "Dune-Fixing in Madagascar: A Line in the Sand for Extreme Weather," World Food Programme, December 17, 2021, www.wfp.org/stories/dune-fixing-madagascar-line-sand-extreme-weather. Accessed 29 January 2025.

Shakespeare, William. *Macbeth*. The Folger Shakespeare, edited by Barbara A. Mowat, Paul Werstine, Michael Poston, and Rebecca Niles. Folger Shakespeare Library. www.folger.edu/explore/shakespeares-works/macbeth/read/. Accessed 29 July 2025.

Part III

Indigenous Nature and Culture

How have Indigenous cultural narratives about nature and the environment been transformed or appropriated? What has the legacy of colonialism and white settler culture meant for the health of the planet?

These chapters all examine the competing narratives of colonialism/white settlers vs. Indigenous cultures in relation to specific geographical and environmental locations—in the Caribbean, in Brazil, in Montana. While each chapter focuses on a very different kind of adaptation—twentieth-century novel adaptation of Shakespeare, television adaptation of narratives of the American West, graffiti adaptation of a Portuguese colonial historical monument—they all demonstrate that Indigenous cultures have had a very different relationship with the environment than their colonizers. Together, these chapters suggest that the study of ecoadaptation in relation to Indigenous cultures is a fruitful area of inquiry with global significance.

9

Indigo's Challenge to *The Tempest*: Explorations in Ecophenomenology

Elizabeth D. Gruber

Introducing Ecophenomenology

Affinities between adaptation studies and ecocriticism are ripe for amplification, given their shared capacity to illuminate the meanings of humanness. Ecologists such as Paul Shepard prefer to define humans as apex predators, highlighting our participation in nature's rhythmic cycles; but we are also a mind-reading species dedicated to symbol-using and storytelling.[1] Of course, our storytelling imperative is a customized biological/evolutionary development, an inbuilt capacity that encapsulates humans' responsiveness to the environment. As phenomenologist Thomas Metzinger explains, "there is an outside world… and

[1] Shepard identifies hunting/carnivorousness as hallmarks of humanness. Mind-reading (variously Theory of Mind) refers to humans' prolific efforts to gauge others' motives or intentions, just as we do with fictional characters. See Lisa Zunshine on the cognitive turn in criticism for examples.

E. D. Gruber (✉)
Commonwealth University of Pennsylvania, Bloomsburg, PA, USA
e-mail: egruber@commonwealthu.edu

P. Demory (ed.), *Ecoadaptation*, Palgrave Studies in Adaptation and Visual Culture,
https://doi.org/10.1007/978-3-032-09847-4_9

an objective reality, but in moving through this world, we constantly apply unconscious filter mechanisms, and in doing so, we unknowingly construct our own individual world," which he calls our "reality tunnel" or "ego tunnel" (9). Thus, the ego tunnel is essentially a story, one that originates in the space where self and world converge. With these points in mind, my essay explores the promising nexus of adaptation studies and ecocriticism by proposing an *ecophenomenology*, an updated version of selfhood that mates a robust or restless interiority with a biologized awareness that emphasizes humans' embedment in the world.[2] Ecophenomenology proves especially useful for assessing Marina Warner's *Indigo, or Mapping the Waters* (1992) as an adaptation of Shakespeare's *The Tempest* (1611).[3] Read in tandem, these texts invigorate ecocritical conceptions of selfhood.

Indigo as Rebel-Text

Currently, many critics view *The Tempest* as an Ur-text of colonialism. Ania Loomba cites the play when theorizing colonialism's related assaults on "physical territories," "social terrains," and "human identities" (182). These intersecting concepts showcase *The Tempest*'s unique significance for ecocritical work, so vital for assessing, then re-making, the human/nature relationship. Similarly, in their study of postcolonial ecocriticism, Graham Huggan and Helen Tiffen identify *The Tempest* as one of "three paradigmatic fictions of colonial encounter" (172). Nevertheless, *The Tempest* masks certain historical/material realities. While Shakespeare nods to burgeoning colonialism and the adjacent slave trade by emphasizing the African heritage of Sycorax and enforced servitude of her

---

[2] The storytelling self compels an ecophenomenological approach, rather than an ecopsychological reading. Nevertheless, ecopsychology's close attention to the issue of how/where to establish the boundary between self and world is relevant. James Hillman outlines the basics of ecopsychology.

[3] The date refers to the first noted performance of the play. I use "adaptation" as an umbrella term, the overarching category enfolding texts that remake identifiable sources, for a variety of distinct aims. Julie Sanders deftly assesses the relevant terms. Likewise, Sujata Iyengar's compendious study of Shakespearean adaptation lays out diverse aims or approaches of this robust assortment of texts.

son Caliban, the play's backstory, featuring Sycorax's mysterious crime and Caliban's attempted rape of Miranda, provides cover for oppressive dynamics. *The Tempest* also effaces nature, unmaking its own origin-story, by transforming the account of a violent storm into a statement of art's surpassing power.[4]

Ranging across several centuries (roughly 1500 to 2000 CE), *Indigo* exposes material conditions submerged in *The Tempest*, activating a transgressive dialogism that emphasizes humans' embedment in the surrounding environment.[5] The novel awakens the mésalliance of the slave trade and colonialism, interlocking systems that proceeded along-side the systematic redefinition of nature as a resource endlessly available for human use.[6] *Indigo*'s reformist energies coalesce in its reimagining of Sycorax: she reappears as an Indigenous Caribbean woman who witnesses the destruction of her way of life. In a playful jab at *The Tempest*, *Indigo*'s Sycorax tells Ariel (freshly conceived as a young Caribbean girl), "I'm considered a famous witch, you know" (109). Here, Warner simultaneously acknowledges and loosens the narrative straitjacket that confines Shakespeare's Sycorax.[7] Warner's flexible third-person narration consistently offers intimate access to diverse characters' viewpoints, extending *The Tempest*'s explorations of selfhood.

Notably, the play advances an early articulation of nature and nurture as co-creators of humanness. The phrase appears in a mini-rant by Prospero, wherein he describes Caliban as "A devil, a born devil, on whose nature / Nurture can never stick" (4.1.188–189). This complaint betrays the crude mechanics of othering: a flawed nature stubbornly resists putatively civilizing instruction. But Prospero's assessment of Caliban side-steps the ecophenomenological awareness intimated in the play,

---

[4] Historian Hobson Woodward meticulously chronicles the actual shipwreck that inspired *The Tempest*.

[5] Arthur Stinchcombe's history of the slave trade and sugar plantations informs my analysis.

[6] My summary draws on Carolyn Merchant's influential environmental history, as well as Todd Borlik's balanced assessment of ecocritical responses to Merchant. Helen C. Scott's Marxist study of *The Tempest* recovers the play's historical implications.

[7] Deborah Willis's skillfully charts early-modern apprehensions of witchcraft, illuminating the presumption that witches coopted/corrupted the maternal role, such as by nursing their "familiars." Also see Lynn Bruckner's account of the era's assumptions about the affinities between women and nature.

whereby the island-environment differentially impacts its inhabitants.[8] *Pace* Prospero, the self is fluid, an entity forged via the ceaseless dance of nature and nurture.

If Prospero now rather easily slips into the category of villain (e.g., emissary of imperialism), his own backstory brims with trauma, as he was betrayed and dispossessed by his brother Antonio. Effects of this treachery linger, evidenced in Prospero's "beating mind," an unruly interiority that frequently pulls him out of the present moment (4.1.163). According to psychologist Cathy Caruth, trauma is essentially a recurring, unwieldy story: it hijacks consciousness and forces the traumatized to keep reexperiencing a painful event (200–204). Chronicling Prospero's mindset, Shakespeare presciently conveys this looping effect, underscoring the centrality of storytelling to humanness.

Indigo builds on this realization. The novel begins in *medias res*, with the Sycorax-analogue, Serafine, entertaining her charges by telling them a story. A Caribbean woman who emigrated to England, Serafine has cared for succeeding generations of children from the Everard family, descended from the very one that steals Sycorax's island in the novel's earlier setting. Serafine's stories create a spiraling effect, knotting past and present realities. For instance, Serafine captivates Miranda Everard with her account of an island whose original inhabitants "'used to live in houses like baskets. Hanging up from the branches of high trees,'" as Sycorax did, before being dispossessed (25). Serafine's story preserves a kernel of history, but she understands the pressure to sanitize certain narratives, given that they are "not seemly for little English girls" (220). Ultimately, however, *Indigo* compels engagement with the messy, painful realities distorted in *The Tempest*. To do so, *Indigo* leans into the inner lives of multiple characters, tracking their customized efforts to navigate the world. We begin with Sycorax.

[8] Mary Thomas Crane writes lucidly of phenomenological issues pertaining to *The Tempest*.

Redeeming Sycorax

Shakespeare's Sycorax never appears in the play, though traces of her presence, her witchy malevolence, linger. Prospero condemns Sycorax's "earthy and abhorred commands," carried out by her "more potent ministers" and fueled by "unmitigable rage" (1.2.274, 276, 277). He imagines the dead woman in tune with nature, possessed of a formidable, earth-bound power. These complaints amount to psychological projection, given that Prospero himself seeks mastery over nature, relies on Ariel to carry out his commands, and battles unruly emotions. Reflecting on Sycorax's confinement of Ariel in a "cloven pine," Prospero reminds his servant, "It was mine art /... that made gape / The pine and let thee out" (1.2.278; 1.2.292–94). As Prospero tells it, when he laid claim to the island and its inhabitants, he was acting as liberator, a bestower of freedom. This self-justifying narrative works because it contrasts Prospero's ostensibly benevolent power with the shadowy menace of Sycorax. In sum, Shakespeare's depiction of Sycorax trades in paradox: she is absent *and* overdetermined, a convenient screen for projecting fears of demonic others.

Indigo steps into the void, creating a satisfying inner life for Sycorax, one grounded in her awareness of participating in a vastness that predates and outlasts humans. Distilling this eco-vision, the narrator reports, "the [I]ndigenous islanders could conceive differently of the time and space they occupied, and see it as a churn or a bowl, in which substances and essences were tumbled and mixed, always returning, now emerging into personal form, now submerged into the mass in the continuous present tense of existence" (129). This description captures the intimacy of bodies and the spaces they inhabit, conceptualizing humans as part of a teeming ecosystem. This friable self nestles comfortably within current ecocritical and posthumanist trends.[9] But Warner also nods to the desire for claiming a "personal form," no matter the ephemerality of

[9] Generally, ecological discourse endorses the friable self. For instance, consider Arne Naess's description of humans as "knots in the biospherical net" (95). Ecofeminists often emphasize humans' membership in an expansive community, but they sometimes leave room for the individuated self. Wendy Donner and Freya Matthew exemplify this stance, so they influence my reading.

this individuated self. Along these lines, the narrator reports, "Unlike her crop," named in the title, Sycorax periodically "needed to lose connection with the island... and when she wanted a clear head to look back into the past and forward into the future, she needed to abandon the continual commitment to the task of survival" (102). Sycorax appreciates the blessings/curses made available through retrospection and anticipation because she understands the demands of organismic life. Nature is thus not irrelevant to cognition but the catalyst for cultivating an inwardness unique to humans.[10] Manifesting this dynamic, *Indigo* emphasizes humans' embedment in the world and makes the case for an ecophenomenological awareness.

Indigo's reparative depiction of Sycorax makes her a fitting rival of Prospero; the novel presents her as "the source of knowledge and the shaper of the order by which" her community lived, in equitable relations with the rest of nature (125). *Indigo* also dislodges the myth of the Dark Maternal, the angst-ridden, witch-mongering assumptions cast onto Shakespeare's Sycorax.[11] Although Warner's Sycorax has given birth, we only see her mothering two adopted children, a son (Dulé, or Caliban) and daughter (Ariel). Sycorax first meets Dulé/Caliban when she senses a macabre flotilla of corpses nearing her island. The murder victims were Africans who had been trafficked into slavery, then cast into the sea. Dule's pregnant mother was one of the victims. Somehow, Sycorax hears the voices of the dead—or near-dead, in Dulé's case. Sycorax ends up delivering the boy from his deceased mother via a daring Caesarean section (95). A whiff of mythic grandeur (salubrious witchiness?) infuses this scene, which celebrates Sycorax's power to summon a child from ironic entombment in his mother's inert body. Here, Sycorax becomes the liberator, taking over the role Shakespeare's Prospero ascribes to himself. Rather than auguring malevolence, this Sycorax's heightened attunement to the world (including its other inhabitants) proves life-sustaining.

---

[10] For useful analyses of consciousness/cognition as evolutionary endowments, see Hugo Mercier and Dan Sperber.

[11] Marina Warner's work on myths and fairy tales offers some good insights into witches and mothers (or witchy mothers), in the tradition that enfolds Sycorax. See especially *Fantastic Metamorphoses, Other Worlds: Ways of Telling the Self*.

Finally, Warner's Sycorax deliberately weaves herself into the world. In a detail exported from Shakespeare, Sycorax's island is described as "full of noises," and she is identified as "the source of many of them" (89). Here as throughout, Warner exploits the alchemizing properties of adaptation, whereby specific passages (damning judgements) from *The Tempest* accrue different meanings. *Indigo* resurrects the aural magic of the play, when the narrator explains that Sycorax "speaks in the noises that fall from the mouth of the wind. It's a way of holding on to what was once hers, to pour herself out through fissures in the rock, to exhale from the caked mud bed of the island's rivers in the dry season" (89). This detail conjures a pleasing image of embedment, of self fully harmonized with world, so it anticipates crucial trends in eco-theory.[12] But this lyrical assessment of Sycorax's power, her felt presence across the island, also foreshadows displacement, with impending loss encoded in the reference to "what was *once* hers."

When a group of Europeans led by Kit Everard arrives on Sycorax's island, they set ablaze the tree in which she had made her home. Sycorax recovers from the fire, but it permanently brands her, so that the usurping Englishmen see her as a "savage and deformed crone," frustratingly oblivious of how their brutality transformed the woman (145). Sycorax survives this encounter, but she never recoups her rightful place on the island. Still, her influence remains, sometimes channeled through the chthonian realm where she had often felt most at home. Specifically, when her body was lowered into the ground, her head remained "nearest to the surface," her mouth adjacent to "the earth and the living who walked on it" (201). The eerie, inevitable process of de-materialization ensues, as Sycorax melds with her island. This post-mortem detail offers a new perspective on Sycorax's ghostly presence in *The Tempest*. In recuperating Sycorax, Warner distills the palimpsestic magic of adaptation, revealing a story whose outlines were just barely visible in *The Tempest*.

As the revitalization of Sycorax indicates, the sustained play between *The Tempest* and *Indigo* revisits familiar dualities—nature/nurture; colonizer/oppressed; self/world; mind/body—in order to advance an updated

[12] Stacy Alaimo's work on trans-corporeality and Karen Barad's account of posthumanist performativity flesh out ecocritical/ecofeminist implications of the embedded self.

version of the human, one that harmonizes our fundamentally biologically selves with the self-reflection and storytelling exclusive to our species. Of course, the stories enabling selfhood are sometimes burdened by woeful distortions, as if to literalize the concept of an ego *tunnel*. *Indigo* exposes this problem by highlighting the story-making of Prospero-analogue Kit Everard, who views events through the warped lens provided by his imperialistic ambitions.

Dueling Narratives: Reimagining Prospero, Caliban, and Ariel

Colonization inevitably generates clashing narratives.[13] Though *Indigo* meticulously chronicles the brutality perpetrated against Africans and Indigenous people, Kit apprehends his actions in grandiose terms, imagining that he and his fellows have "fought off the tribe who were famous cannibals and befriended the Indians who weren't… and put them to work" (112). He marvels that some of the Indigenous islanders were "good workers, all green fingers and thumbs, with a yeoman's knowledge of the properties of the ground, of water and seed and drainage" (112). This patronizing judgment, characteristic of colonialist rhetoric, implicitly divides experiential knowledge from ostensibly "higher" intellectual pursuits.[14] Kit exalts his own actions, vowing, "I shall found a garden in these western isles," as if he undertaking a quasi-divine mission to create a new Eden (153). Adrift in self-deluding mythopoeia, Kit fails to recognize his own savagery, though it damages the world he claims as his own.

Kit's sense of proprietorship over the island underwrites his interactions with Dulé, their dynamic generating a new version of the Prospero/Caliban relationship. In a letter Kit writes to his patron, he frankly discloses the planned enslavement of Dulé, who will be a "bondsman,"

[13] Edward Said's *Culture and Imperialism* remains an excellent resource for assessing the function of those narratives generated by colonialism.

[14] Huggan and Tiffen document this point in *Postcolonial Ecocriticism*. Likewise, Loomba addresses the "constant negotiation" between "colonialist knowledge" and "indigenous ideas" in her theorization of postcolonialism (80).

permanently "hobbled" and under vigilant surveillance (199). Kit adds that Dulé "has already learnt how to curse," and that some of his men call him "cannibal," though he himself opts for "the lisping usage of the children, Caliban" (199). Here, again, *Indigo* confronts *The Tempest*, compelling awareness of its founding assumptions. Thus, while the play invites readers/playgoers to discern the wild Other—the "cannibal"—in Caliban, the novel exposes the cynical linguistic games, the re-naming campaigns, marshalled in support of unjust sociopolitical arrangements. Furthermore, *Indigo* ironizes the wilding of Caliban. If Kit fails to recognize the full humanity of Dulé, the novel imagines a robust interiority for the latter, who proves more restrained—more compassionate or civilized—than the men who occupy his adopted home. As the narrator explains, when Dulé initially sees the invaders on his island, "It did not come to his mind to murder any of them; it was not his way, or the way he had been raised, to take life in cold blood, without a formal challenge or warning," (158).

Dulé's predicament, a radical dislocation from his heritage, intensifies the significance of the storytelling self, always formed via responsiveness to the environment. The narrator explains, "Dulé developed an idea of the past that was foreign to the people among whom Sycorax had been born and raised; it was a lost country for him, which he wanted to rediscover" (105). Of course, we can only ever encounter the past as an "idea," a narrative that—when we are fortunate—imbues experiences with purpose and meaning. Dulé's forced displacement skewers his reality. He tries to compensate by building a ladder into the sky, which he says reflects the desire to bridge the gap "Between the time now and the time I can't remember" (105). Frustration bristles through this declaration, particularly when we note the centrality of memory to consciousness, hence to the self-concept.[15] Moreover, the image of Dulé trying to climb a ladder to nowhere conveys the disastrous effects of living permanently amid savagely imposed exile. Severed from his land, community, and past, Dulé lacks a proper narrative framework by which to understand his life. This is untenable because, as Charles Taylor

[15] Patrick Colm Hogan makes this point about the interconnections between memory/consciousness/identity.

comments, "doing without frameworks is impossible" for humans (27). Frameworks, or stories, generate the possibility of community, and they also enable the self, that is, a secure vantage point from which to engage the world.

Kit's brutality extends to the Indigenous people he encounters. Shortly after igniting Sycorax's arboreal home, Kit fixates on Ariel, who was already traumatized by watching her mother burn. As Kit rapes Ariel, she experiences a radical self-division. The narrator explains that there were "Two Ariels, one outside the other, each watching the other, curious, inert" (152). The mind becomes a refuge, enabling Ariel to detach from the violence. In defiant or edgy fashion, Ariel chooses to define the assault as evidence of her own power, proof that she is capable of "filling a man with desire, just as Sycorax had always" predicted (152). Obviously, this narrative fails to comport with current apprehensions of sexual violence. But we should note how it furthers Warner's critique of colonizing rhetoric by inverting the associations of mind and colonizer, body and Indigenous subject. That is, Ariel recognizes Kit as a prisoner of his bodily appetites, whereas she retreats into the infinite expanses of her mind.[16] Kit, however, remains convinced of his moral superiority over the woman whom he consistently views as primitive or savage, as if projecting onto Ariel the violence of his own desires—just as Shakespeare's Prospero does with Caliban.[17]

Highlighting the interior, story-making capacities of Dulé, Sycorax, and Ariel, *Indigo* redresses the othering at work in *The Tempest*, which brackets characters into groups and builds walls between them. Reflecting on politicized identities in Shakespeare, Ambereen Dadabhoy identifies a "pernicious belief in the difference, cultural and racialized," of colonized subjects as a staple feature of what she calls "imperialist technologies" (229). Dadabhoy, however, rejects the possibility of a "shared human condition," espousals of which she deems the result of an abiding, if historically unexamined, ethnocentrism that dominates

---

[16] Huggan and Tiffen address "the Cartesian separation of mind and body" as one of the crucial "technologies of 'othering'" endemic to colonial encounters (172).

[17] *Indigo*'s shrewd complication of "primitive" traits exemplifies the incisive arguments presented by Marianna Torgovnick in her comprehensive account of the uses/representations of "primitive" or "savage" tropes.

Shakespeare scholarship. I agree that we must continue to challenge *pseudo*-universals, which pretend to inclusivity while reserving humanness for certain subjects. But it might be the opportune moment to posit a retooled universality, a shared human nature, made available through the aesthetic realm.[18] For all its imperfections, *The Tempest* nods in this direction. *Indigo* completes this theme.

Epilogue: Revisiting Art and Nature

Shakespeare's Prospero orchestrates various spectacles, his art evidently able to contravene natural laws. This theme crescendos when Prospero has Ariel stage a false banquet to entice/entrap Alonso, Antonio, and Sebastian. When these three approach the sumptuously laid table, the meal disappears, and they find themselves immobilized, forced to turn inward and reckon with the knowledge of their past transgressions (3.1.53–109). In this interactive playlet, spectators become performers, unwitting participants in a living (if temporarily frozen) tableau. Of course, the transforming magic only really works on Alonso, who suddenly, devastatingly, struggles against the weight of his own guilt. He believes the natural world sings of his transgressions, noting the effect "is monstrous, monstrous!" (3.3.95). Prospero forces Alonso to confront a formidable karma, namely, that his misdeeds have cost him his son, who "i' th' ooze is bedded" (3.3.100). Enlisting natural elements to achieve his desired aesthetic effect, Prospero engineers Alonso's psychic revolution.

To complete its redefinition of humanness, *Indigo* channels its predecessor's metadramatic extravagance. More specifically, the novel features a performance of *The Tempest*, delivered before a late-twentieth-century audience in England. The narrator highlights the reaction of Miranda Everard, Kit's daughter, as she watches the action unfold. This version of the play directly confronts Caliban's enslavement. Miranda sees that Caliban is immobilized "because he had his ankles tied together; she then saw him thrash out with his tethered limbs and lumber, like a stranded sea creature, toward the actress" playing Miranda (367). This

[18] See monographs by Ellen Dissanayake and Denis Dutton for analyses of art-as-instinct.

searing image fuses the novel's past and present sequences, recalling the moment when Sycorax and her fellow islanders first witness the bilboes, or shackles, binding the ankles of the Africans who had been trafficked into slavery. Additionally, this scene reassesses (Shakespeare's) Caliban's evident disruption of ontological categories, as when Trinculo cannot decide whether Caliban is "a man or a fish," (2.2.24–25). In the adaptation, the dehumanizing effects of subjugation account for Caliban's ostensible strangeness. Enacting the tangible horrors of slavery, *Indigo* critiques the dormancy of this motif in *The Tempest*.

The play-going Miranda watches the on-stage Miranda assert dominance over Caliban, mocking him and condemning him as a "slave" who resists "any print of goodness" (367). The narrator comments, "The young woman [playing Miranda] was ranting the speech, she was taunting him with smiles, with seductive looks which then turned to scorn, and she was wiping tears from her eyes in horror at the memory of the ingratitude he'd shown her care for him. While Miranda, watching, listening, shivered" (367). Even the spectator-Miranda's body responds to the brutal scene, a piece of stage-craft at once faithful to and critical of Shakespeare's play, because it refuses to ennoble or even to rationalize the naked subjugation of one person by another. *Indigo*'s Miranda finds herself riveted by the actor playing Caliban. She "felt the edges of her body flare as if she were against the light and outlined in the sun's fire" (367). Quickly, however, "Miranda turned in fury on herself," scolding, "'You're trapped in the fantasy, that someone like him could melt you and take you down to the thing you've lost touch with—the longed-for, missing Primitive. Light my fire! Show me paradise! Blow me away! Be an animal, show me the beast inside'" (368). Miranda turns inward, assessing the seemingly automatic associations of the Black actor with the "Primitive."

Returning to Metzinger's explanation of the ego tunnel, Miranda recognizes the "unconscious filters" that predicate her responses to Caliban. Distilling this process, *Indigo* captures art's potency: it intervenes on a neural level, disrupting commonplace perceptions, associations lodged in memory and automatically activated as we move through the world. When it comes to interpersonal dynamics, this means that we unconsciously respond to those we perceive as different in pre-patterned ways, as dictated by the "mental lexicon" or "semantic memory," patterns of association that influence our relations with one another and the rest of nature.[19] *Indigo* confirms that bias, including racial prejudice, resides in the brain, but it is formed, reinforced, or challenged by sociocultural experiences and environments. Art has the potential to sponsor meaningful, lasting change, precisely because of its capacity to reform or update the lexical entries. *Indigo* effectively turns the force of *The Tempest* back on itself, calling upon the transporting/transforming effects of art to indict injustice and barbarism. Activating the magic of the aesthetic realm, Warner shows how art has woven into the story of humanness. It originates out of an inborn capacity—true of any ability we might name—yet is deeply imbricated in the cultural and sociopolitical forms that lend shape and meaning to experience. As *Indigo*'s Miranda recognizes, art sometimes knocks down barriers that otherwise seem impenetrable.

As it happens, Miranda had met the actor playing Caliban, Shaka Ifetabe (formerly George Felix), years earlier, but the awkwardness of this encounter left them exiled to each other. Sharing *The Tempest*, albeit as spectator and actor, occasions the start of Miranda and Shaka's relationship. The narrator explains, "They had begun play. Their openings were well-tried, unadventurous. But these same familiar moves would take them in deep: face-to-face and piece by piece they would engage with each other so raptly that for a time they would never even notice anyone else looking in on the work they were absorbed in" (373). This marks another strategic nod to *The Tempest*, reprising the image of Miranda and Ferdinand playing chess as they await departure from the island. But in having Miranda choose *Caliban* (or the actor who plays this role), and vice versa, *Indigo* offers an alternative to *The Tempest*. In the play, the betrothal of Miranda and Ferdinand, aided if not instigated by Prospero, ensures that power replicates itself, and the ruling community remains homogeneous. *Indigo* counters with a model of inclusivity or expanded community. Such transformations necessarily manifest at the macro-level, but they originate from the psychological work Miranda undertakes. Her willingness to assess and update her personalized story,

[19] Hogan outlines these cognitive processes and concepts.

her ego tunnel, holds the key to felicitous change. In sum, *Indigo*'s interventionist approach to *The Tempest* tracks fantasies of unbreachable difference, but the novel ultimately sets forth an alternate vision, premised on shared humanity.

Accordingly, *Indigo*'s taut engagement with *The Tempest* distills the possibilities for aligning ecocriticism and adaptation. The joint consideration of biological/evolutionary developments and the primacy of storytelling holds exciting promise, pointing toward a fresh, viable take on universality. As I hope to have shown, ecophenomenology tracks this embodied/embedded self.

Works Cited

Alaimo, Stacy. "Trans-Corporeal Feminisms and the Ethical Space of Nature." *Material Feminisms*, edited by Stacy Alaimo and Susan Hekman, Indiana UP, 2008, pp. 237–64.

Barad, Karen. "Posthumanist Performativity: Toward an Understanding of How Matter Comes to Matter. *Material Feminisms*, edited by Stacy Alaimo and Susan Hekman, Indiana UP, 2008, pp. 120–54.

Borlik, Todd. *Ecocriticism and Early Modern Literature: Green Pastures*. Routledge, 2011.

Bruckner, Lynne Dickson. "N/nature and the Difference 'She' Makes. *Ecofeminist Approaches to Early Modernity*, edited by Jennifer Munroe and Rebecca Laroche, Palgrave MacMillan, 2011, pp. 15–35.

Caruth, Cathy. "Trauma and Experience." *Theories of Memory: A Reader*, edited by Michael Rossington and Anne Whitehead, Johns Hopkins UP, 2007, pp. 199–205.

Crane, Mary Thomas. *Shakespeare's Brain: Reading with Cognitive Theory*. Princeton UP, 2001.

Dadabhoy, Ambereen. "The Unbearable Whiteness of Being (in) Shakespeare." *Postmedieval: A Journal of Medieval Cultural Studies* 11, no. 2–3: 228–235, 2020.

Dissanayake, Ellen. *Art and Intimacy: How the Arts Began*. U of Washington P, 2000.

Donner, Wendy. "Self and Community in Environmental Ethics." *Ecofeminism: Women, Culture, Nature*, edited by Karen J. Warren, Indiana UP, 1997, pp. 375–89.

Dutton, Denis. *The Art Instinct: Beauty, Pleasure, and Human Evolution.* Bloomsbury Press, 2009.

Hillman, James. "A Psyche the Size of the Earth: A Psychological Foreword." *Ecopsychology: Restoring the Earth, Healing the Mined*, edited by Theodore Roszak, Mary E. Gomes, and Allen D. Kanner, Sierra Club Books, 1995. pp. xvii–xx.

Hogan, Patrick Colm. *Cognitive Science, Literature, and the Arts: A Guide for Humanists*. Routledge, 2003.

Huggan, Graham and Helen Tiffin. *Postcolonial Ecocriticism: Literature, Animals, Environment*, 2nd ed., Routledge, 2015.

Iyengar, Sujata. *Shakespeare and Adaptation Theory*. The Arden Shakespeare, Routledge: 2023.

Loomba, Ania. *Colonialism / Postcolonialism: The New Critical Idiom.* 3rd edition, Routledge, 2015.

Matthews, Freya. "Ecofeminism and Deep Ecology." *Key Concepts in Critical Theory: Ecology*, edited by Carolyn Merchant, Humanities Books, 1999, pp. 235–45.

Merchant, Carolyn. The Death of Nature: Women, Ecology, and the Scientific Revolution. Harper & Row, 1980.

Metzinger, Thomas. *The Ego Tunnel: The Science of the Mind and the Myth of the Self*. Basic Books, 2009.

Naess, Arne. "The Shallow and the Deep. Long-Range Ecology Movement: A Summary." *Inquiry*, vol. 16, 1973, pp. 95–100.

Said, Edward. *Culture and Imperialism.* Vintage Books, 2024.

Sanders, Julie. *Adaptation and Appropriation*, 2nd ed. Routledge, 2016.

Scott, Helen C. *Shakespeare's* Tempest *and Capitalism: The Storm of History.* Routledge, 2021.

Shakespeare, William. *The Tempest. The Oxford Shakespeare: The Complete Works.* 2nd, ed., edited by John Jowett, William Montgomery, Gary Taylor, and Stanley Wells. Clarendon Press, Oxford, 2005. 1223–43.

Shepard, Paul. *The Others: How Animals Made Us Human.* Island Press, 1996.

Stinchcombe, Arthur L. *Sugar Island Slavery in the Age of Enlightenment: The Political Economy of the Caribbean World*. Princeton UP, 1995.

Taylor, Charles. *Sources of the Self: The Making of the Modern Identity*. Harvard UP, 1989.

Torgovnick, Marianna. *Gone Primitive: Savage Intellects, Modern Lives*. The U of Chicago P, 1990.
Warner, Marina. *Fantastic Metamorphoses, Other Worlds: Ways of Telling the Self*. Oxford UP, 2002.
Warner, Marina. *Indigo, Or Mapping the Waters*. Simon & Schuster, 1992.
Willis, Deborah. *Malevolent Nurture: Witch-Hunting and Maternal Power in Early Modern England*. Cornell UP, 1995.
Woodward, Hobson. *A Brave Vessel: The True Tale of the Castaways Who Rescued Jamestown and Inspired Shakespeare's* The Tempest. Viking, 2009.
Zunshine, Lisa. *Why We Read Fiction: Theory of Mind and the Novel*. Ohio State UP, 2006.

10

"A Huge Mass in a Single Hand": *Yellowstone* and the Selling of Montana

Katie Kane

"This is America, we don't share land here."
—John Dutton, *Yellowstone*

One of the many commodity tie-ins associated with the proliferating and popular television series *Yellowstone* (2018–2024)—an object range that includes ranch jackets, cowboy coffee, and photo-embossed "character" pillows, among a great many other things—is the Paramount Plus–sponsored Yellowstone Monopoly board game™.[1] If commodity tie-ins

[1] The wide range of commodities associated with *Yellowstone* also includes the products placed in the show itself, such as Kimes Ranch Apparel, Coors beverages, the Buffalo trace line of liquor, Dodge Ram trucks, YETI coolers, Carhartt, Tito's Vodka, 6666 Grit and Glory Beverages (a company owned by series creater Taylor Sheridan), Stetson, etc. See Swystum, Guttadaro, for more details. For the official commodity tie-in website of the series, see the official *Yellowstone Shop*, where as of February 2025 the shopper is encouraged to get ready for "The Return of 1923."

K. Kane (✉)
University of Montana, Missoula, MT, USA

P. Demory (ed.), *Ecoadaptation*, Palgrave Studies in Adaptation and Visual Culture,
https://doi.org/10.1007/978-3-032-09847-4_10

Fig. 10.1 Yellowstone Monopoly Board. Photo courtesy of the author

to any film or television show "adapt" a concept, visual theme, or character for circulation in the marketplace, then the Yellowstone Monopoly board game adapts and extends the logic defining the governing conceit of the show: the concept of monopoly itself—the cornering and domination of an economic sphere by a unitary and single incorporated body (Fig. 10.1).

On the screen and on the board, *Yellowstone* narratively adapts and refigures colonialist monopolies by depicting a concentration in the means of beef production in southwest Montana and a centralization in the key zone of its marketplace: grazeland (property). This monopoly over grazeland and cattle by one individual or family reanimates and reinstalls the central story or ideology of settler ranching in Montana: the notion of a Western individual achieving personal apotheosis in dominating the wilderness of the "frontier." These material and ideological counterparts of the profit-driven and accumulative American West define the overarching life philosophy of John Dutton (Kevin Costner), the patriarch of the Dutton family and owner of a 30,000-acre ranch in Paradise Valley, Montana, known in the show as "The Yellowstone."

Dutton's struggle to control his land monopoly drives the plot of the series, which circulates almost exclusively around Dutton's understanding that land is both material and cultural power: "Leverage is knowing that if someone had all the money in the world, this [the Dutton ranch land] is what they'd buy" ("Daybreak"). "Leverage" is just another word for monopoly power in closing out competition for the sake of dominating the market, but in the context of the television series leverage is also what allows for a cornering of the marketplace of mythic ideas of the West as articulated both in practice and in cultural documents.[2]

In the American West, there is no more significant model of the concentration of economic and social power than the large ranch, as its role as a fiction foundational to the settlement of the interior of the continent makes clear. Two examples clarify the importance of the monopolistic ranch in the settlement of the American West—one archetype is fictional and one autobiographical, but both are equally fabulated and importantly part of the narrative of territorial and economic conquest in the West. *Giant*, Edna Ferber's 1952 novel, translated onto the Hollywood screen in 1957, is an exemplary manifestation of the Western epic tradition that lionizes the extraordinarily wealthy and monopolistic ranch owners. The novel, the film, and its subsequent successor, *Yellowstone*, glorify and celebrate—albeit in complicated soap opera conventions—the owners of the monopoly: Bick Benedict and, much later, John Dutton.

The recent adaptation of the North Dakota history of Theodore Roosevelt into a short-lived television series, *Elkhorn* (2024), makes clear the closed circuit of exchange that "fiction" and "fact" have in the cycle of adaptation that undergirds most Western genres in which art and life imitate and mutually reinforce each other.[3] No matter that Roosevelt only briefly established and ran his ranch in western North Dakota in the 1880s and that it was never a going concern; the penetration

[2] For a useful discussion of the role of leverage in the construction of monopolies see Blair.

[3] In recognition of the popularity of neo–ranch westerns following the success of *Yellowstone*, the INSP Network produced and broadcast *Elkhorn* about the establishment of Theodore Roosevelt's ranch in Little Missouri Country for one season in 2024. INSP is a self-identified socially conservative "family entertainment" network that reaches more than sixty million households.

of his ideas about the West and its "perfect freedom" circulate everywhere in Western ideology—endlessly recycling Roosevelt's romance of the "strenuous life."[4] According to Roosevelt, true individual independence was more easily achieved for an American man on a cattle ranch than elsewhere: "I do not believe there ever was any life more attractive to a vigorous young fellow than life on a cattle ranch in those days. It was a fine, healthy life, too; it taught a man self-reliance, hardihood, and the value of instant decision" (Roosevelt 21*)*. From Roosevelt to Benedict to Dutton, Western "heroes," fictional and semifictional, secure their personal power by establishing and protecting land empires that concentrate capital and property.

This tendency of capital to concentrate into monopolies is perhaps most sharply put by Karl Marx in volume I of *Capital:* "Capital grows in one place to a huge mass in a single hand, because it has in another place been lost by many" (777).[5] In the case of ranches and farms in both North Dakota and Montana (where the historical reality and imaginary of the ranch property looms large), there is an emerging trend to concentrate and centralize into the hands of the super-rich property that was once rural, public, and small ranch land. In the current post–COVID-19 pandemic period, and in light of global and national wealth concentration, land empires resulting from property consolidation and concentration have significantly shifted the region's geographies. According to Kathleen Epstein et al., it is high net worth owners who are the "leading buyers of agricultural land" in the Greater Yellowstone Ecosystem (GYE)—a region that is "rich in landscape amenities" and "global conservation value" (432). The impact of this monopoly control is manifold, involving the elimination of small family farms and ranches, the acceleration of amenity migration into the area, the "dude-ification" or tourist transformation of the region, and loss of biodiversity in crops

[4] See Slotkin, who identifies the centrality of Roosevelt's personal version of the myth of the frontier to popular culture and the development and circulation of the western literary and filmic genres.

[5] Marx does not directly discuss the creation of monopolies that are largely the result of inherent contradictions in the system that result in stagnation and lack or loss of growth; he does, however, note in blunt terms the terminal tendency of capitalism to concentrate in the hands of a few.

and livestock. But most central to the "Yellowstone Effect" is the way in which monopolization of land accelerates ecological damage.

Most ecological work in the American West prior to the second decade of this century focuses on land fragmentation as a central threat to sustainability and biodiversity, but recently scholars and activists have begun to acknowledge the specific and growing threat that land monopolies pose in the region. As Julia H. Haggerty et al. argue in "Rural Land Concentration and Protected Areas: Recent Trends from Montana and Greater Yellowstone," private land use "determines the quality and quantity of wildlife habitat and the provisioning of ecosystem services" (693).[6] In a data-rich and deeply researched study of the concentration of land ownership from 2005 to 2018 in the Greater Yellowstone Ecosystem, Haggerty et al. document the ways in which monopolization of the land is well underway. For instance, the Greater Yellowstone and Northern Great Plains areas studied by Haggerty experienced an increase in land ownership concentration between 2005 and 2018, as the "total number of landowners holding properties of 640 acres or more declined by 6.4%, from 2,993 to 2,801," while the average size of these large holdings correspondingly grew by 7% to 4,141 acres in 2018 (695). In the twelve counties studied, the expansion of extra-large properties was particularly striking, with the forty-eight largest holdings in twelve counties increasing by nearly 235,000 acres, 70% of which were absorbed by the largest properties. In a further striking development in some of the studied counties, the largest landholdings doubled in size, and the disparity between the first- and second-largest properties increased fivefold. In addition to documenting what amounts to a massive land grab, Haggerty et al. describe the important economic reality behind the monopolization of land in the region: the concentration in land ownership is largely driven by new investment; as in four of the five counties where the gap between the two largest holdings increased, new landowners entered the market and were responsible for consolidating land through multiple acquisitions.

---

[6] The Haggerty article is a valuable resource for scholars looking at ecology and land concentration. It is a "first in the scholarly literature" that studies "trends in land concentration near protected areas" in the context of the 24.9 million acres that comprise the Greater Yellowstone and Northern Great Plains ecosystems.

This land monopoly has serious and damaging implications for the migration corridors and critical habitats of western Montana. Amenity migrants and absentee landlords use the land they accrue to create investment portfolios and affective geographies for the sake of heritage lifestyles and "ranch aesthetics."[7] In "With, Not For, Money: Ranch Management Trajectories of the Super-Rich in Greater Yellowstone," Kathleen Epstein and her colleagues explore the centrality of aesthetics and heritage in large ranch ownership in part via a case study of conceptions of tradition and ownership on The Stonefly, a smaller ranch on the Wyoming side of the Greater Yellowstone Ecosystem. The essay records the investment of the owner of The Stonefly in notions of "tradition":

> There is a lot about this place that's all about traditions.... Partly in jest but partly seriously [I have] referred to [the Stonefly] as a *lifestyle museum* because it's not only the fact that we have all of these great historic buildings and historic furnishings and stuff like that. It's also the way the ranch is used. It's quite traditional. (437, emphasis mine)

If aesthetic and affective concerns result in landscape petrification—whereby land becomes the mere container for family and personal history—material matters are equally at stake in the transformation of landscape following human designs and desire. In order, for instance, to guarantee that hobby fishing matches the expectations generated by Norman Mclean's semi-autobiographic novel, *A River Runs Through It* (1976), rivers and streams are often diverted or surrounded by riprap to ensure their "stability." Furthermore, as monopoly ranches seek to expand amenity assets specific to the region—trout and elk, first among them—similar attempts at stabilization and control characterize their private and often idiosyncratic systems of elk management. Fluctuation in elk population can significantly impact both the biome and animal communities since elk waste, also known as nitrogen deposition, can alter the soil, plant diversity, and vegetative growth. There may be a future in

---

[7] Hjerpe et al. define "amenity migration" as "the relocation for quality of life purposes as opposed to monetary enhancement, [which] has been occurring for decades and has been particularly pronounced in the American West where the phenomenon peaked in the 1990s. Researchers have illustrated that some places are more attractive for migrants than others and that certain regional amenities hold considerable influence on where migrants relocate" (56).

which the Greater Yellowstone Ecosystem and Montana might accommodate public land use, agricultural and livestock production, and land monopolies, but the most likely imminent reality seems to be a consolidation of private enclosures in which the terms of development and interaction with surrounding ecosystems are defined by elite individuals. Most ominous of the impacts of monopoly ranches on the surrounding ecosystems and migrating wildlife is the rise of largely enclosed and elite-dominated ownership patterns with "a mix of HNW [High Net Worth Individuals] and traditional landowners" (Epstein 688).

The ideology of private ownership and land concentration permeates the *Yellowstone* series, but that discourse and practice are most visible and compressed in the legion of cultural memes or axioms generated by the series and taken up by fans. The "wisdom of *Yellowstone*"—to borrow a designation that circulates widely in contemporary popular culture—is a kind of comforting cracker-barrel philosophy, just one horse stall over from the "common sense" so frequently invoked in the American West to justify a host of material and cultural practices, including settlement itself: "You're either born a willow or born an oak," "I'd rather kill a thousand men than shoot another horse," "I don't make the way," "Be a man about it, don't scream," "I wish that they'd come back to life so I could kill 'em again," etc. This series of pithy, easily commodified maxims is ready-made to grace a million T-shirts and coffee mugs, and the collective message concerns the fundamental value of land monopoly (Figs. 10.2 and 10.3).

Yellowstone as a whole promotes and fosters this ideology of private ownership and land concentration, but nowhere is this distillation of the show's philosophy more clear than in the two self-designated "iconic" and core declarations of the Dutton family: "protect the land" and "we don't choose the way."[8] The sentiment "protect the land" is widely distributed in *Yellowstone* but always uttered in the context of other concerns such as "family," "ownership," "heritage," or the marketplace. In the official trailer for the third season of the series John Dutton states: "I made a promise to protect it [the land]... for me and my

[8] See the official *Yellowstone Shop* for the ways in which objects and mottos are routinely designated as "iconic."

Fig. 10.2 "Protect the Land" Coffee Mug. Photo courtesy of the author

family: can't break it." Here, in recounting the promise he made to his father, Dutton articulates the relationship of blood and soil. And in the context of "blood heritage," violence is appropriate to protect property: "You protect what is yours by any means necessary" ("Behind Us"). It is perhaps an understatement to note that spectacular violence and bloodshed is the Dutton family's preferred way of "protecting" the land, a set of actions presented as both de facto and appropriate. In the fourth episode of season 3, "Going Back to Cali," for instance, Dutton explains to the cartoonish bikers from California, "Man said if you didn't leave, we'd bury you here: You didn't leave." Sometimes in *Yellowstone* the land seems the pretext for a larger goal of violence, but land is always to be owned, to be added to, and to be defended in the only arena that

Fig. 10.3 "We Don't Choose the Way" T-Shirt. Photo courtesy of the author

matters: the marketplace. This, indeed, is "the way" that the show relentlessly touts: "It's the one constant in life. You build something worth having, someone's gonna try to take it" ("Enemies"). During a conversation about labor, masculinity, and respect, among other things central to land monopolies, Rip Wheeler (ranch foreman) and Kayce Dutton (son of the owner of the means of production) speak to the structure underpinning the Dutton land empire:

Rip: You let me handle the wranglers ... that's my job.
Kayce: Alright, get them up to pasture nine. You choose the way.
Rip: I don't choose the way. I make sure nobody questions yours. ("The Reek").

In like manner, in season 1, episode 1, the oldest of John Dutton's sons, Lee Dutton, says to his younger brother, Jamie, "we don't make the way, little brother," as ranch hands and Livestock Association agents leave the ranch to confront the cattle-rustling members of the wholly fictitious Native nation, "The Confederated Tribes of Broken Rock," led by Thomas Rainwater. In fact, Dutton's attempt to win a battle he is fighting with Dan Jenkins and Rainwater, who are attempting to build a hotel and casino adjacent to Dutton's property, structures a crucial narrative arc in the first season of the show and underscores vigilante violence as necessary in the service of "the way" (as well as of his ethic of "protect the land"): he ultimately dynamites a river to keep the plan from fruition. Here "the way" is clearly articulated by John Dutton who says, when marching out to battle with his rifle in his hand, "Everyone's forgotten who runs this valley" ("Kill"). In a collision of monopoly capital and the ethic of settler colonialism—particularly in the casting of Indigenous people as a threat to resource consolidation—the Duttons reassert their primary and exclusive right to "run the valley."

In all this running of things in the region by folks like the Duttons, Montana as an idea continues to depend upon the process of primitive accumulation under the terms of white settler colonialism. As Yellowknives Dene First Nation scholar Glen Coulthard explains, settler colonialism depends upon the reproduction of the means of domination of Indian Country "structured into a relatively secure or sedimented set of hierarchical social relations that continue to facilitate the dispossession of Indigenous peoples of their lands and self-determining authority" (7). As a central cultural part of a set of linked social relations, *Yellowstone* is part of this dispossession by accumulation.[9] Capitalist settler colonialism in Montana and elsewhere in the West is perpetually extractive, forever located in the phase of primitive (primary) accumulation wherein dispossession of Indigenous land, resources, personhood, and sovereignty is an ongoing and constitutive feature of the reproduction of colonial

[9] Here I am using and slightly retooling David Harvey's concept of "dispossession by accumulation," as articulated in *The New Imperialism*, among other places. Harvey updates Marx by explaining the ways in which colonial-capitalist states and economies shift away from the overt violence of bloody territorial expropriation into the consolidation work of state policy and the promulgation of culturally enforced ideology.

and capitalist social relations in an age often designated as "postcolonial." This notion of primitive accumulation extends in Montana from the practice of energy extraction often located in or near Indigenous land, to the concentration of land in the hands of capital-flush developers, to the disproportionate incarceration of Indigenous men and, in larger numbers, women, in Montana's prisons and jails. Organizing the regions of "The West" into a zone, as such, has always been a matter of the transformation of the region's territory into its interdependent and correspondent commodities, from the material to the ideological. An ideological and cultural machine for this process of transformation, *Yellowstone*—with its narrative recruitment of the history of Native nations in Montana in the service of its ideological project—distorts the history of land tenure and sovereignty in a narrative of seamless, ahistorical land transfer from Indigenous peoples to white settlers.[10]

By series end, it transpires that the Duttons have—with convenient permission from a tribal elder—only been borrowing their land in Paradise Valley. Unlike the owner of a monopoly ranch, the owners of The Yellowstone are, have been, and it appears will continue to be "stewards" of the land, as explained in the last episode of season 5 by the late series rival to John Dutton as main character, his great-aunt, Elsa Dutton:

> One hundred and forty-one years ago my father was told of this valley and here's where we stayed. Seven generations. My father was told they would come for this land, and he promised to return it. Nowhere was that promise written. It faded with my father's death but somehow lived in the spirit of this place. Men cannot truly own wild land. ("Life is a Promise")

While the Duttons have been giving a pretty good imitation of monopoly in land ownership, what with the not "sharing" land and violently responding to any perceived or imaginary threat, the series does

[10] Cherokee scholar Liza Black makes clear the crimes of appropriation and ideology that the series commits: "*Yellowstone* spins a simple story about white America.... But *Yellowstone*'s subtext is another thing entirely: the settler-colonial version of American history, which offers didactics on human nature rather than confronting the history of Native peoples."

its best in the late episodes of the fifth season to forward an alternative narrative of land tenure that excludes the history of monopolistic accumulation and white settler triumph. The ranch land was not appropriated: in *1883*, the prequel and origin story of the Dutton family, James, the family patriarch, is directed to the location of "Paradise Valley" in southwestern Montana by the wise Crow elder, Spotted Eagle (who also directs healing ceremonies for Elsa). This is the well-worn white fantasy of recognition by Indigenous peoples—"we are one of the good ones"—who, having performed the necessary nomination of their colonizers, remain adjunctive to the settler narrative and who quickly vanish once the structural need is achieved.

Spotted Eagle:	I know a place for you. You go through that pass, then you follow the river south. I used to hunt that valley as a boy. The winters are cruel, but the summers are rich and a man who plans can thrive. And you look like a man who plans.
James Dutton:	What's the valley called?
Spotted Eagle:	I don't know the word in your language, but it's, uh, when you die, you go there.
James Dutton:	Heaven.
Spotted Eagle:	No, there's another word, that's not it.
James Dutton:	Paradise.
Spotted Eagle:	Yes, Paradise, good name. But you know this: that in seven generations, my people will rise up and take it back from you.
James Dutton:	In seven generations, you can have it.

By the fifth season of *Yellowstone*, this post-fact narrative arc imported from 1883 comes into play and makes clear the ways in which settler colonialism in Montana cannot be understood as appropriating land from Native nations for the sake of the extraction of social and monetary profit. The Duttons, like all other settlers for whom their family is metonym, have simply been borrowing the land, caring for it, in order to return it, in seven generations, to the Broken Rock community. They are giving the land back, of course, once it has become economically unsustainable for the Dutton family to maintain its monopolistic ranch—a form of "get-out-of-jail free" card in the terms of the game of Monopoly.

Yellowstone's exculpatory "land-back" plot device is achieved in metaphysical terms that hinge on visions and prophecies that use Native traditions—at least as imagined by Taylor Sheridan, writer and co-producer of both *Yellowstone* and *1883*—to center white characters and values. For instance, this invocation of the "seven-generation" prophecy parodies the Haudenosaunee principle of future-oriented behavior in the present specifically with regard to resources.[11]

In the closing montage of the *1883* series, which in its narrative structure ties up the plot of *Yellowstone*, Elsa, again the authoritative voice of the Dutton family, reminds the viewers that the perfect world ("the world of you") is made of "dreams and memories," and it is always morning in America, the "sunrise" of beginning, and not the sunset, horizon-collapsing world of the United States in the early twenty-first century ("This is Not"). This concept of a land of endless new beginnings where none of the wages of environmental and community destruction will ever come due is, of course, one of the central struts of the American Dream. Bad behavior in the present—the extraction of coal near the Apsáalooke reservation, the drilling for oil on or near Niitsitapi and Assiniboine Country, the rise in crime and missing and murdered Indigenous women and people near sites of energy extraction, as examples—will never come home to roost in this imagination of the American Dream. It can all just be given back, as "promised" in the revisionist history of US settler colonialism.

With its contributions to the construction of monopolies of power, land, and capital, it is apt that among the many, many pieces of leisure ephemera generated for *Yellowstone* by Paramount and for sale in its online shop—ephemera such as golf balls, cornhole games, hoodies, hats, and shirts (in either flannel or camo to suit your imaginative desires

[11] Oren Lyons, the Haudenosaunee Faithkeeper of the Wolf Clan of the Onondaga Nation and the Seneca Nation of the Six Nations of the Grand River spoke about the seventh generation principle in his 1977 oration "Where is the Eagle Seat?" given before the UN: "A thousand years ago we were given this message by the Creator; we were given a government by the Creator. This government was not manufactured from the minds of men, it was given to us; and we were to cherish it. And each generation was to raise its chiefs and to look out for the welfare of the seventh generation to come." In *Yellowstone*, the seven generations become merely a historical interval emptied of its weighty ethical implications for the present day.

Fig. 10.4 *Yellowstone* Monopoly Cards. Photo courtesy of the author

to hunt or to kill), fire pit tables, sweatshirts, and so on—that Yellowstone Monopoly is one of the "best sellers" of the *Yellowstone* collection[12] (Fig. 10.4).

To state the obvious: it is, in fact, as the cover of the game advertises, a "Fast Dealing Property Trading Game"—based on luck and force—that is at the heart of Montana's settler colonial history:

> All's fair in love and land in this ultimate game for Yellowstone fans! **MONOPOLY®: Yellowstone** lets players buy, sell, and trade well-known locations anchored by the Dutton family's coveted cattle ranch, including Kayce and Monica's House and the Painted Horse Casino & Hotel. Gear up with collectible cowhand-related tokens and post Wranglers and Foremen to secure your grounds to be the last operator standing with custom *Yellowstone* currency to win!" (*Yellowstone Shop*)

This might also be said of the Missoula housing market as well as of the history of settler colonialism in Montana, regardless of which historical era one is speaking.

No present-day phenomenon demonstrates the direct impact of *Yellowstone* and its adaptation of the foundational practices and fictions

[12] In the *Yellowstone* Monopoly edition, the "Community Chest" cards have been updated with Rip Dutton's line, "We Don't Choose the Way." The "Chance" cards have been rebranded as "Protect The Land" cards.

of settler colonialism that it underwrites than the post–COVID-19-pandemic population boom Montana has experienced since 2020. A study by Patrick Barkey and Melissa Weddell for the University of Montana's Bureau of Business and Economic Research (BBR) makes clear the impact of *Yellowstone* on tourism in the state in the post-pandemic period. They found that 17.4% of visitors polled for the study list *Yellowstone* as a motivating factor for their visit. In 2021 alone this influence translated to approximately 2.13 million visits. This influx of visitors has coincided with a broader demographic shift in Montana. Between 2020 and 2022, according to the *2024 Labor Day Report*, the state experienced a surge in in-migration, with over 40,000 more people moving in than leaving, marking one of the highest net migration rates in the country (Montana Department of Labor and Industry). While multiple factors contributed to this trend, the increased national visibility of Montana, partly due to *Yellowstone*, may have played a role in attracting new residents in addition to tourists.

In a rather stunning way, then, recent demographic movement to Montana owes a good bit of its momentum to the show *Yellowstone*, which both gathers and advances the mythology of the West, a narrative that repeats and circles its own discourse and history—as tragedy and farce at once. *Yellowstone* and its associated commodities produce a portable, fungible "Montana" commodity that can be moved anywhere in both the imaginative and ideological space of the United States, as frontier, and as predicated on territory, on land (the solidness melting into air in the process of the production and consumption of the series). Even though the axiom of the game is to protect the land, the method of "conservation" follows Lockean notions of property ownership and transforms the land into a site for the production of deeds and the emplacement of houses and hotels. Notions of leisure and hierarchy (class and status) are embedded in the game, both historical and experiential. And in this board game version of our lives in the West, to "be the last operator standing with custom Yellowstone currency [is] to win!" To extend the lexicon of the game a bit further without breaking the hermeneutic: We don't make the way here in Montana or elsewhere. "The way" of land monopoly in Montana is both a historical legacy of a colonial past and the ethos of a present defined by the extension of that settler logic into a twenty-first-century land grab.

Yellowstone, the television series, is one of the most important and critically neglected aspects of the material and ideological marketing of Montana in the post-COVID period—as a space of white flight and as a homeland for "political refugees" from blue states. The series attempts to re-install and to rewrite the originary terms of North American settler colonialism in its narrative reversal of the relations between white settlers and Native nations, in its valorization of an ideology of "rugged individualism" aligned with an ersatz ranching-style capitalism, and in its ethically empty "environmental" ethos contained in the maxim "protect the land," among other things. Moreover, as a narrative adaptation of a once waning but recently recrudescent story of the American West that positions the land as "beautiful" without considering its extreme environmental precarity and its long history with extractive economies, including mining, logging, agriculture, and ranching, *Yellowstone* perpetuates dangerous fantasies of a frontier utopia where the application of individual will and "hard work" are necessary to achieve colonial domination and ownership of the land and, by extension, the western version of the "American Dream." In particular, the ways that *Yellowstone* depicts issues such as water rights conflicts, the Endangered Species Act, wildlife compensation funds, stream access and western fence laws, brucellosis transmission, and, of course, the nature of property rights contribute to a revanchist notion of Western environments as the "last best place" for anti-environmental ideologies and practices. *Yellowstone*, then, is an adaptation of an older and familiar set of narrative and ideological coordinates—those longitudinal and latitudinal lines that chart the map of the settler colonial frontier of the "West."

References

Barkey, Patrick, and Melissa Waddell. "Assessing the Impact of the *Yellowstone* TV Series on Montana's Tourism Economy." *Montana Business Quarterly*, January 2023, pp. 1–30.

"Behind Us Only Grey." *Yellowstone,* season 2, episode 8, written by Taylor Sheridan, John Linson, and Brett Conrad, directed by Taylor Sheridan, Paramount, 14 August 2019.

Black, Liza. "On *Yellowstone* and the White Desire to Control the Narrative." *High Country News,* April 22, 2022, www.hcn.org/issues/54-5/indigenous-affairs-art-on-yellowstone-and-the-white-desire-to-control-the-narrative/. Accessed 22 July 2025.

Blair, R. D., and Esquibel, A. K. "Some Remarks on Monopoly Leveraging." *The Antitrust Bulletin,* vol. 40, no. 2, 1995, pp. 371–396. https://doi.org/10.1177/0003603X9504000204.

Coulthard, Glen Sean. *Red Skin, White Masks: Rejecting the Colonial Politics of Recognition.* U of Minnesota P, 2014.

"Daybreak." *Yellowstone,* season 1, episode 1, written by Taylor Sheridan and John Linson, directed by Taylor Sheridan, Paramount, 18 June 2018.

"Enemies by Monday." *Yellowstone,* season 2, episode 9, written by Taylor Sheridan, John Linson, and Eric Jay Beck, directed by Guy Ferland, Paramount, 21 August 2019.

Epstein, Kathleen, et al. "With, Not for, Money: Ranch Management Trajectories of the Super-Rich in Greater Yellowstone." *Annals of the American Association of Geographers,* vol. 112, no. 2, Feb. 2022, pp. 432–48.

"Going Back to Cali." *Yellowstone,* season 3, episode 4, written by Taylor Sheridan and John Linson, directed by John Dahl, Paramount, 12 July 2020.

Guttadaro, Andrew. "A Tribute to Yellowstone's Many Jackets," *The Ringer,* July 2018, www.theringer.com/2018/07/12/tv/kevin-costner-yellowstone-jackets. Accessed 22 July 2025.

Haggerty, Julia H., et al. "Rural Land Concentration & Protected Areas: Recent Trends from Montana and Greater Yellowstone." *Society & Natural Resources,* vol. 35, no. 6, June 2022, pp. 692–700.

Harvey, David. *The New Imperialism,* Oxford UP, 2003.

Hjerpe, Evan, Anwar Hussain, and Thomas Holmes. "Amenity Migration and Public Lands: Rise of the Protected Areas." *Environmental Management,* vol. 66, no. 1, 2020, pp. 56–71.

"Kill the Messenger." *Yellowstone,* season 1, episode 2, written by Taylor Sheridan and John Linson, directed by Taylor Sheridan, Paramount, 27 June 2018.

"Life is a Promise." *Yellowstone,* season 5, episode 14, written by Taylor Sheridan and John Linson, directed by Taylor Sheridan, Paramount, 15 Dec 2024.

Linson, John, and Taylor Sheridan, creators. *Yellowstone.* Paramount, 2018–2024.

Lyons, Oren. *An Iroquoian Perspective. Learning to Listen to the Land*, edited by W. B. Willers, Island Press, 1991.

Marx, Karl. *Capital*, Volume I. Penguin, 1990.

"A Monster is among Us." *Yellowstone,* season 1, episode 7, written by Taylor Sheridan and John Linson, directed by Taylor Sheridan, Paramount, 8 August 2018.

Montana Department of Labor and Industry. *2024 Labor Day Report. Montana* Government Publishing Office, 2 Sept. 2024, https://lmi.mt.gov/_docs/Publications/LMI-Pubs/Labor-Market-Publications/LDR24_Final.pdf. Accessed 22 July 2025.

"The Reek of Desperation." *Yellowstone,* season 1, episode 7, written by Taylor Sheridan and John Linson, directed by Stephen Kay, Paramount, 10 July 2019.

Roosevelt, Theodore, and Frederic Remington. *Ranch Life and the Hunting-Trail.* Century Co, 1888.

Sheridan, Taylor, creator. *1883.* Paramount+, 2021–2022.

Slotkin, Richard. *Regeneration Through Violence: The Mythology of the American Frontier*, 1600–1860. Wesleyan University Press, 1973.

Swystum, Jeff. "Yellowstone is Brimming with Brands," *Medium* Jan 17, 2022, jeffswystun.medium.com/yellowstone-is-brimming-with-brands-b188410792d6. Accessed 22 July 2025.

"This is Not Your Heaven." *1883,* season 1, episode 10, written by Taylor Sheridan, directed by Ben Richardson, Paramount+, 27 Feb 2022.

Yellowstone Shop. Spike Cable Networks, Inc., SnowCommerce, Inc., 2025, yellowstonetvshop.com. Accessed February 2025.

11

"This Is Indigenous Land": Adapting and Protesting Colonialist Discourses through Graffiti in Forte Dos Reis Magos, Brazil

Augustto Correa Cipriani and Clara Matheus Nogueira

On September 7, 2023, the 201st anniversary of Brazil's Independence, the walls of Forte dos Reis Magos (Three Wise Men Fortress) were painted with vibrant red dye, displaying the phrases "*Aqui é terra indígena*" ("This is Indigenous land") and "*Não ao PL/2903*" ("No to PL/2903"). Located in Natal, Rio Grande do Norte (RN), the fortress is a major symbol of the city, dating back to its foundation in the late sixteenth century. Forte dos Reis Magos is deeply entrenched in the history of colonization in the region, as the Portuguese colonizers built it to protect the lands they were claiming—both from Indigenous peoples and French invaders. The Forte's graffiti points to the beginning of the military dictatorship in Brazil, in 1964, when protesters began targeting

A. C. Cipriani (✉)
Universidade Federal de Minas Gerais, Belo Horizonte, Brazil
e-mail: acipriani@ufmg.br

C. M. Nogueira
Universidade Federal de Campina Grande, Campina Grande, Brazil
e-mail: clara.matheus@professor.ufcg.edu.br

P. Demory (ed.), *Ecoadaptation*, Palgrave Studies in Adaptation and Visual Culture,
https://doi.org/10.1007/978-3-032-09847-4_11

specific buildings with their political graffiti, adapting the narratives associated with historic buildings according to their political demands.

During the month the graffiti occurred, PL 2903—a bill addressing the recognition, demarcation, use, and management of Indigenous lands—was voted on by the Federal Senate of Brazil. At the same time, one of its most controversial legal theses, the Time Frame, was under scrutiny by the Brazilian Federal Supreme Court. According to this thesis, only lands occupied by Indigenous peoples on October 5, 1988—the date Brazil's constitutional text was enacted—could be legally demarcated. Critics argue that this framework disregards the historical displacement of Indigenous peoples, many of whom had already been expelled from their ancestral lands by 1988. The bill further threatened Indigenous rights by permitting public authorities to construct roads and other infrastructure on Indigenous lands and by weakening protections for isolated Indigenous communities. Described as "Legislated Genocide" by The Articulation of Indigenous Peoples of Brazil (5), PL 2903 posed a significant threat to both Indigenous rights and environmental preservation. While 29% of the land surrounding Indigenous territories is deforested, deforestation within Indigenous lands amounts to only 2% (APIB 18), a difference that can be traced to Indigenous beliefs about the natural world, and highlighting the critical role of the demarcation of Indigenous land for the biosphere's protection. On September 21, the Brazilian Federal Supreme Court declared the Time Frame legal thesis unconstitutional. Nevertheless, on December 28, PL 2903/2023 was enacted into law as 14.701. At the time of writing (December, 2024), the validity of the Time Frame thesis is again under debate by the Brazilian Federal Supreme Court, with significant implications for Indigenous rights and environmental protection (Fig. 11.1).

Forte dos Reis Magos is a landmark that attracts visitors interested in Brazil's colonial history. Built in 1598, the fortress served as the starting point for the settlement of Natal, functioning as a base for Portuguese colonizers to expel French traders who were extracting brazilwood in the region and had already established contact with the Potiguara people (Cascudo 22). Shaped like a star, the white fortress now houses historical artifacts, including the Marco Colonial de Touros—one of the first

Fig. 11.1 Forte dos Reis Magos west wall covered in graffiti

Portuguese *padrão*[1] in Brazil—as well as archaeological finds such as Indigenous ceramic pots. These artifacts highlight the presence of Indigenous peoples during the early years of the fortress. The permanent exhibit features informative panels that reveal a clash between progressive and colonial perspectives on Brazilian history, particularly regarding the representation of Indigenous peoples.

The graffiti painted on the Forte's walls was monumental in scale, covering the entire west wall, which is approximately sixty meters wide. Three distinct graffiti dominated the space: "*Não ao PL / 2903,*" which occupied more than half of the wall, featured letters and numbers that extended from the ground to the battlement, rendered in a simplistic style; "*Aqui é terra indígena,*" a more stylized graffito, occupied about a quarter of the wall; and in the southwest corner, Indigenous symbols and motifs were prominently displayed. The scale of the inscriptions suggests that they were likely created by a group of coordinated graffiti artists, during nighttime hours. The graffiti adapted the Forte into a canvas, but it also adapted *pichação*, the most notable form of Brazilian graffiti. The Forte's graffiti abridged the distance between contemporary graffiti and the political graffiti that was preeminent during the early days of the Brazilian military dictatorship. The history of Brazilian graffiti in the twentieth century can be summarized in three major trends: a first wave of political graffiti, which served as an outlet for dissidence during the early years of military dictatorship; a second wave, in the decades of 1970 and 1980, was characterized by the emergence of poetic graffiti, mostly by teenagers and university students, who disrupted the urban semiosis with enigmatic phrases; and finally, the rise of *pixo* in the late twentieth century, influenced by American hip-hop culture and rock album imagery, which became the most renowned form of Brazilian graffiti (Lassala 82, Cipriani 70). All these iterations of graffiti fall under the umbrella term *pichação*, derived from the Portuguese word for tar

[1] *Padrãos* were stone pillars that symbolized the discovery and conquest of territories during Portuguese maritime exploration in the fifteenth and sixteenth centuries. Typically, they were adorned with the coat of arms of Portugal, a cross, and other official inscriptions. When discussing the early colonization of the state of Rio Grande do Norte, Câmara Cascudo compares the construction of Forte dos Reis Magos to the placement of a *padrão* in a new territory, as the Forte marked the founding of the city of Natal (26).

("*piche*"), as they are understood as acts of defacement. In contrast, even though *pichação* is a form of graffiti, in Brazil the terms "graffiti" or "*grafite*" typically refer to a specific form of urban art that is aesthetically closer to mural painting or "pieces"—short for masterpieces.

In the twenty-first century, graffiti and *pichação* have been increasingly polarized. In Brazil, the differentiation made between graffiti and *pichação* is framed through a play of dualities, ranging from oppositional categories—such as commissioned versus illegal—to more radical contrasts such as beauty versus ugliness, art versus vandalism, revitalization versus defacement, and finally, image versus word. This final contrast is particularly significant as *pichação* is among the few contemporary artistic practices in which manual writing survives: in *pixo,* individual or collective tags are written on walls and buildings as signatures claiming visibility in the urban landscape. When nomenclature is considered, the term "graffiti" has been lexically dominated by authorized practices, leading to the exclusion of illicit and transgressive forms of graffiti from the category, even among their practitioners. Gustavo Lassala demonstrates such a dichotomy through "Graffiti workshops": "There are Graffiti workshops that encourage *pixadores* or taggers to acquire the Graffiti's ideals and techniques. This practice reveals the current trend to 'domesticate' taggers so that their work become[s] socially acceptable, which is not the case with pixo" (39). Within the urban context of Brazil, there is a prevailing tendency to associate colorful imagery—often regarded as aesthetically pleasing—with graffiti, while written words, whether political or otherwise, are frequently perceived as visually aggressive *pichação.*

The graffiti at Forte dos Reis Magos cannot be categorized only as *pichação*; rather, it constitutes political graffiti. When arguing for the political dimension of graffiti, Waldner and Dobratz compare graffiti to the other forms of contentious political participation, noting their shared nonnormative and nonconventional intentions: "Graffiti, squatting, and culture jamming are all about taking over space and expressing counter hegemonic views. All are attempts to 'take over' and replace current dominant discourse" (381). For the authors, graffiti may represent an act of appropriation in which the graffitists claim or dispute the ownership over the spaces they intervene in—whether a wall, a building, a neighborhood, etc. Waldner and Dobratz do not comprehend graffiti as inherently political, unlike Hanauer (306), who understands graffiti as a transgressive and anti-official literacy act of marginalized groups. The graffiti in Forte dos Reis Magos, nevertheless, is undeniably political as it engages with an ongoing legislative and judiciary issue.

On the one hand, the reference to PL 2903, "This is Indigenous land," is, as stated, directly tied to a ruling by the Supreme Court. On the other hand, it is an adaptation of a recurring slogan in Indigenous people's protests across Brazil. "*Brasil é terra indígena* [Brazil is Indigenous land]" serves as a rallying cry in demonstrations against the Time Frame thesis, often accompanied by localized variants, such as "*São Paulo is Indigenous land*." This rallying cry has been adapted in various ways: in 2023, for instance, it was used as the name for an Indigenous music festival funded by the Brazilian government. The decision to write the deictic "*aqui* [this, or here]" at the Forte walls is thus remarkable. It echoes the broader assertion that all Brazilian territories are originally Indigenous lands, while also drawing attention to the current striving for Indigenous land demarcation in the state of Rio Grande do Norte. Furthermore, it symbolically reclaims the fortress constructed by the Portuguese in the late sixteenth century as Indigenous. Although there is currently no federally recognized Indigenous land in Rio Grande do Norte, numerous villages and communities continue to assert their rights to both Indigenous identity and land demarcation.

By framing the Forte dos Reis Magos as Indigenous land, the graffiti also encourages a reexamination of the fortress's history and its permanent exhibits. The history of Indigenous peoples in Rio Grande do Norte is characterized by the erasure of their presence, particularly from the nineteenth century until the late twentieth century (Cavignac and Alveal 34). The permanent exhibit at Forte dos Reis Magos stands in contrast to this historical erasure. There is a room dedicated to a panoramic view of the history of Indigenous peoples in Brazil presented in banners alongside an assortment of illustrative objects. Many texts featured in this room are drawn from the *Guia Cultural Indígena Rio Grande do Norte* [Rio Grande do Norte's Indigenous Cultural Guide], edited by Julie Cavignac and Carmen Alveal. The exhibition features a QR code that leads visitors to the guide, which documents the Indigenous presence in the region, compiles historical information about these communities, and highlights tourist, educational, and gastronomic attractions. Remarkably, the image that illustrates Cavignac and Alveal's guide in the exhibit is a stylized map of the state, adorned with Indigenous motifs, which reads "*RN é terra indígena* [RN is Indigenous land"]. This image was present in the exhibit before September 7, 2023, and remained on display in late 2024, nearly a year after the graffiti removal. Another striking feature of the exhibit is the centrality of Filipe Camarão and Clara Camarão. Filipe Camarão was a member of the Potiguara people who was raised by Jesuit missionaries, from whom he learned Portuguese and Latin and was baptized as a Catholic. He later married Clara, who was also of Potiguar origin. In traditional Brazilian historiography, Filipe Camarão has been portrayed as a protonational hero for his military contributions against the Dutch invasions in the seventeenth century. At the Forte exhibit, Filipe and Clara Camarão are celebrated for their military prowess while their Indigenous heritage is not explicitly emphasized. Furthermore, their histories are not included in the space dedicated to Indigenous peoples in Rio Grande do Norte, but are instead presented in a separate room. This spatial separation reflects a broader tension between two distinct narratives: one that evidences the violent erasure of Indigenous identities in the region, and another that valorizes assimilated Indigenous peoples and emphasizes colonial accomplishments. This tension illustrates the fortress's dual role as both a site for critically reexamining institutional perspectives on Indigenous peoples in Brazil and a symbol of colonial power.

Through adaptation, the graffiti further stresses this narrative tension. The political graffiti at Forte dos Reis Magos adapts Indigenous peoples' imagery and motifs, such as the last and stylized "a" in the word "*indígena,*" and the shade of red used evokes *urucum* paint, a red dye made from seeds traditionally used in body paint. The motifs painted in the corner of the wall resemble both "Potiguara resistance" and "coral snake," as described by Falcão (192, 282) and Pereira (111), as well as the "butterfly" motifs following Bezerra's (36) depiction. In a news article in the left-wing newspaper *A Verdade*, Îawotĩ Potygûara Tarairiý asserts that the icons on Forte dos Reis Magos' graffiti represent a local variation of the butterfly motif, which has been a symbol of resistance since the colonial period for the Potyguara de Potyguasú people. The struggles of the Indigenous peoples in Brazil are further highlighted by the residue of the graffiti's red paint, which dripped onto the ground in several places, resembling blood spatters. The spattering serves as a blunt reminder of the genocidal history of ethnic cleansing that persisted for centuries in the Americas, while simultaneously giving prominence to the resilience of Indigenous peoples who continue to struggle and fight for survival to this day. The Forte's graffiti appropriates the building and adapts it. Declaring Forte dos Reis Magos as Indigenous land transforms it into a site of Indigenous reclamation, countering its historical role as one of the starting points of Portuguese settlement. The Forte's graffiti appropriates the techniques of *pixação* and adapts it. Infusing *pixação* with Indigenous imagery, the graffiti asserts Indigenous identity, presence, and resistance within a space historically dominated by colonial narratives (Figs. 11.2 and 11.3).

In our proposal to interpret graffiti as an act of adaptation, we highlight its potential to transform the meaning of the wall it adorns. Following Kamilla Elliott, we argue for adaptation as an "interactive, relational process that changes entities to suit new environments; it is also a term describing an entity thus changed" (198). In this sense, the graffiti adapted the Forte itself, transforming the dominant colonial narrative associated with the building to suit the context of the demonstrations against the Time Frame thesis. Understood as a form of "exposed writing" (Petrucci 88), graffiti disputes official narratives by presenting discursive dissent to a plural and broad audience. Written at a site that monumentalizes the colonial history of the city, the graffiti reveals how the Forte functions as a *lieu de mémoire*, that is, a "site of excess closed upon itself, concentrated in its own name, but also forever open to the full range of its possible significations" (Nora 24). As the graffiti is inseparable from the wall, those who read the inscriptions on the Forte, simultaneously saw the text, the motifs, and the fortress. As Linda Hutcheon puts it, "For the reader, spectator, or listener, adaptation as adaptation is unavoidably a kind of intertextuality *if the receiver is acquainted with the adapted text*" (21). Adapting Hutcheon's stance, if a passerby is familiar with the colonial history of the Forte, the graffiti promotes an unavoidable interrelation between the Indigenous genocide and the ongoing struggles for

Fig. 11.2 Detail of the graffiti at Forte dos Reis Magos

Fig. 11.3 Red paint splatters from the graffiti at Forte dos Reis Magos

land demarcation in Brazil. It cements the absurdity of the Time Frame thesis, as Indigenous history in the region dates back to the construction of the Forte in 1598—a fact supported by evidence displayed in the exhibition. The graffiti sheds an unavoidable light on the arbitrariness of setting 1988 as the cut-off for land demarcation. Writing a political claim, particularly from an Indigenous standpoint, on the walls of an ancient Portuguese fortress is a gesture of defiance aimed at the country's colonial past and present. The graffiti artists understood Forte dos Reis Magos as a "tainted political symbol"; therefore, their act of vandalism not only raises awareness about the Time Frame thesis but also serves as a form of broadcasting counter-speech that undermines the "derogatory pedestalling" of Indigenous peoples and the authority of the state to arbitrate on the shared history of Indigenous and non-Indigenous Brazilians (see Lai 2, 3). The Forte, a white, centenary fortress, was adapted by the red and splattered graffiti into a visual argument against the Time Frame thesis.

Ultimately, since 1998, *pichação* has been considered an environmental crime in Brazil, a profound irony when the graffiti at Forte dos Reis Magos is considered. Originally, the law criminalized all forms of graffiti. More recently, in 2011, a legal adjustment created a statutory distinction between *pichação*, understood as an act of vandalism and defacement, and graffiti, recognized as "artistic manifestation," if authorized, or even commissioned, by property owners or local authorities (Brazil). Although according to the current law the graffiti at Forte dos Reis Magos could be understood as *pichação*, thus an environmental crime, it raised awareness about the Time Frame thesis—one of the most severe threats to Brazil's existing environmental protection measures.

The Indigenous lands perform a critical role in environmental protection. Indigenous philosopher and environmentalist Ailton Krenak explains how the Krenak Nation perceives nature through different lenses when compared to people he calls "city folk." First, the Doce River, on which the Krenak and other Indigenous populations depend, is not considered a resource, but a family member—the river that they call "Watu" is considered their grandfather. In "Tomorrow is Not for Sale," Krenak explores how anthropocentrism puts humankind in antagonism to nature:

> For a long time, we were fed the story that we, humanity, stand apart from the great big organism of Earth, and we began to think of ourselves as one thing, and Earth, another: Humankind versus Earth. We have to abandon our anthropocentrism. There's a lot more to Earth than us, and biodiversity doesn't seem to be missing us at all. Quite the contrary.
> My communion with what we call nature is an experience long scoffed at by city folk. Rather than see any value in it, they poke fun at it: "He talks to trees, he's a tree-hugger; he talks to rivers, he contemplates the mountains." But that's my experience of life. I don't see anything out there that is not nature. Everything is nature. The cosmos is nature. Everything I can think of is nature. (Krenak, ch. 1)

This perception that nature is not something separate from humanity leads to an approach toward the environment that is at odds with that of the government. Krenak sees the politics of demarcating lands for Indigenous peoples, in the twentieth century, as an endeavor rooted in the colonial issue of managing the last surviving traditional communities: "Our reserve, the Krenak reserve, was created in the 1920s basically to function as a container, and they stuffed us inside it. It confines us. We're shut in there and kept out of the way." However, a century later, the consequences of land demarcation became evident, as these lands, according to Krenak, became the "last remaining swaths of land where nature still prospers, where our needs for food and a home can be met."

Indigenous peoples have acted as safeguards for the environment, through their survival and ways of life. As the Yanomami shaman Davi Kopenawa famously puts it, the forests and their people hold back the sky from climate calamity:

> If they [white people] persist in devastating the forest, all the unknown and dangerous beings that inhabit and defend it will take revenge. They will devour them relentlessly with the same voracity with which their epidemic fumes devoured our people. They will burn down their land, smash their houses in storm winds, or drown them in floods of mud and water. All this could truly happen one day if all the remaining shamans die and their orphan *xapiri* [spirits] run away, enraged by their fathers' death. Then there would soon be nothing left in the forest—in what

> white people call "nature"—save the *në wãri* evil beings. (Kopenawa and Albert 404–405)

That is, nature protection is central to the Yanomami cosmovision. Wildfires, hurricanes, floods, and extreme weather conditions are interpreted as nature's revenge for deforestation. Further, Kopenawa declares the shamans as the forest's guardians: they keep control of the forest's spirits. The fight of the surviving Indigenous communities is a fight for environmental preservation—these peoples have been resisting annihilation while battling deforestation. Thus, Krenak claims for a shift in humankind's relation to nature:

> We ought to recognize nature as an immense multitude of forms, each and every piece of ourselves included, for we, too, are part of the whole: 70 percent water, and a host of other minerals. We created this abstract idea that humankind is the measure of things, and we go out into the world like so many little steamrollers, safe in the general assurance that we belong to one big humanity for whom the world was made, and that we are therefore entitled to stomp around taking whatever we want. Engagement with any other possible mindset means hearing, feeling, smelling, inhaling, and exhaling layers of all the stuff that is not "us" and so must be "nature," but which isn't quite nature because it's too much like us. There's something quasi-human about these layers, something we recognize, but which is disappearing from the earth, being exterminated from the interface of totally-human humans. These quasi-humans are a motley crew in their thousands who insist on declining to join this dance of civilization, technology, and planetary control. (Krenak, ch. 1)

The "dance of civilization" that Indigenous communities declined to join brought us to the Anthropocene. Therefore, the vandalism on Forte's walls expresses an ecological duty. The fight for Indigenous rights and nature protection converge.

A few weeks after the graffiti was painted, the fortress was restored to its untainted whiteness. According to local media, convicts, overseen by correctional officers, were responsible for the graffiti's erasure. Visually and structurally, the building returned to its previous state. After all, the materiality of graffiti is nothing but paint. The exhibition hosted at the building does not mention the incident. The authors of the graffiti remain undisclosed. However, as a political act of expression, its effects are lasting. As a *lieu de mémoire*, the Forte became a space where colonial history was reinterpreted and called into question. The graffiti was inscribed onto the history of the building, as a site of contested memory and a testimony to Indigenous peoples' past and present struggles.

Works Cited

Articulation of the Indigenous Peoples of Brazil (APIB). *Setbacks on Indigenous Rights in Brazil*, 2023, apiboficial.org/files/2023/09/marcotemporal_panfleto_en_tela.pdf. Accessed Nov. 2024.

Bezerra, Nilton Xavier, editor. *Artesanato Potiguara: Catu Dos Eleotérios e Sagi Trabanda*. Editora IFRN, 2017.

Brazil, Presidency of Republic, *Lei No 12.408 de 25 de Maio de 2011*, 2011.

Cavignac, Julie A., and Carmen Alveal, editors. *Guia Cultural Indígena Rio Grande Do Norte*. Flor Do Sal, 2019.

Cascudo, Luís da Câmara. *História do Rio Grande do Norte*. 2nd ed., Achiamé, 1984.

Cipriani, Augustto Correa. "Escritas Contemporâneas Brasileiras: Escrita Manual a partir de Guilherme Zarvos e Gustavo Speridião." *Universidade Federal de Minas Gerais*, 2021.

Elliott, Kamila. *Theorizing Adaptation*. Oxford UP, 2020.

Falcão, Emmanuel De Sousa Fernandes. "Grafismo e Discurso Identitário Indígena Potiguara Da Paraíba No Século XXI." *Universidade Federal da Paraíba*, 2022.

Hanauer, David I. "The Discursive Construction of the Separation Wall at Abu Dis." *Journal of Language and Politics*, vol. 10, 2011, pp. 301–21.

Hutcheon, Linda. *A Theory of Adaptation*. Routledge, 2006.

Krenak, Ailton, and Anthony Doyle. *Ideas to Postpone the End of the World*. E-book, Anansi International, 2020.

Kopenawa, Davi, and Bruce Albert. *The Falling Sky: Words of a Yanomami Shaman*. Translated by Nicholas Elliott and Alison Dundy, Harvard UP, 2013.

Lassala, Gustavo. *Pichação Não É Pixação: Uma Introdução À Análise de Expressões Gráficas Urbanas = Picho Is Not Pixo: Introduction to an Analysis of Urban Graphic Expression*. Altamira Editorial, 2017.

Lai, Ten-Herng. "Political Vandalism as Counter-Speech: A Defense of Defacing and Destroying Tainted Monuments." *European Journal of Philosophy*, vol. 28, no. 3, 7 Aug. 2020, pp. 602–616, doi.org/https://doi.org/10.1111/ejop.12573.

Nora, Pierre. "Between Memory and History: Les Lieux de Mémoire." *Representations*, vol. 26, 1989, pp. 7–24. https://doi.org/10.2307/2928520.

Pereira, Erika Danielly Florêncio. "Design e Memória Cultural: Análise Dos Grafismos Corporais Da Etnia Potiguara." *Universidade Federal de Campina Grande*, 2020.

Petrucci, Armando. "Potere, spazi urbani, scritture esposte: proposte ed esempi." *Publications de l'École Française de Rome, Culture et Idéologie Dans La Genèse de l'État Moderne: Actes de La Table Ronde de Rome*, vol. 82, Oct. 1985, pp. 85–97.

Tarairiý, Îawotĩ Potygûara. "Monumento Colonial Em Natal Recebe Manifestação Contra o Marco Temporal." *A Verdade*, 15 Sept. 2023, averdade.org.br/2023/09/rn-decora-forte-dos-reis-magos-em-protesto-contra-o-marco-temporal/. Accessed 12 Dec. 2024.

Waldner, Lisa K., and Betty A. Dobratz. "Graffiti as a form of contentious political participation." *Sociology Compass*, vol. 7, no. 5, 17 Apr. 2013, pp. 377–389, doi.org/https://doi.org/10.1111/soc4.12036.

Part IV

The Nature of Genre Adaptation

How do genres represent and adapt environmental themes? How might genres adapt to explore the very concept of nature?

The chapters in this section explore the way that genres adapt to environmental pressures. The first chapter examines the relationship between climate change and Jim Jarmusch's slow cinema approach to the zombie film; the second shows how environmental changes in the western United States have influenced the Coen brothers' adaptation of Western stories and mythologies in their Netflix series; and the third imagines how changing conceptions of the nature of gender might create an evolutionary shift in the romcom that would subvert the binary oppositions that seem so natural to the genre. In each of these cases, the authors are exploring exceptions to—or possibly mutations of—conventional genre patterns, illustrating the nature of genre adaptation: genres evolve when mutated forms successfully adapt to new environments.

12

The Dead Don't Die and Zombies Don't Run: Adapting the Zombie Apocalypse to Slow Cinema

Agnieszka Rasmus

The zombie is a particularly fertile figure for adaptation, an empty vessel, literally and figuratively, which needs to feed on other living (texts) to survive, continue, and remain relevant. It has evolved and mutated throughout decades, each time resurfacing to comment on social, cultural, and political changes and upheavals, showing its adaptive potential and malleable nature. It has represented our fears of consumerism, nuclear war, contagion, and technological advancement, serving as a "barometer of cultural anxiety" (Dendle 45). This paper explores how the zombie can also be adapted to offer a valid critique of our attitude to the environment, whether as consumers exploiting the planet's resources or passive observers of nature's progressing degradation.

According to Ian Olney, the rise of the zombie figure in the new millennium is not so much rooted in our perception of their difference from us but rather in our recognition that we are fundamentally

A. Rasmus (✉)
Department of Studies in English Drama, Theatre and Film, University of Lodz, Łódź, Poland
e-mail: agnieszka.rasmus@uni.lodz.pl

P. Demory (ed.), *Ecoadaptation*, Palgrave Studies in Adaptation and Visual Culture,
https://doi.org/10.1007/978-3-032-09847-4_12

the same. He also notes how "The living dead have undergone significant revision in twenty-first-century zombie cinema" (82). For one thing, they are now "more likely to sprint than shamble across the screen" (82). The slow-moving creature introduced in George A. Romero's seminal *The Night of the Living Dead* (1968) has been replaced by a frenzied and enraged monster in *28 Days Later* (2002), inspiring other productions to follow, such as *Dawn of the Dead* (2004), *Zombieland* (2009), *World War Z* (2013), *Z Nation* (2014–2018), and *Army of the Dead* (2021). Hordes of violent flesh-eating monsters now typically run toward their victims. This change has invigorated the genre, keeping it "up to speed" with technological advancement and the general trends in action cinema.

The present discussion is focused on Jim Jarmusch's unusual addition to the recent zombie cycle, *The Dead Don't Die* (2019), which is an attempt to appropriate and reconfigure the creature from a fast and angry attacker back to the shuffling and slow figure of Romero's cult classic. In this way, Jarmusch not only returns to the roots of the zombie tradition but also subverts current genre conventions and audience expectations. Using the zombie as a metaphor for our inert response to climate change, he builds a frustratingly quiet and slow apocalypse, punctuated by moments of dead time, whose "heroes" are just as zombified as the meandering and melancholic creatures they silently observe from a distance. Jarmusch's critical response to the firmly established and flourishing cycle shows, as Robert Geal argues, that tracing the development of such crucial cultural texts can help us "address the fundamental ecocritical question—what is it about our culture(s) which cause(s) us to endanger our own survival by treating the biosphere with such contempt?" ("Towards" 6). This essay is an answer to Geal's call to adaptation scholars to look at "how human attitudes towards the environment, and cultural expressions of those attitudes in the form of various kinds of texts, are adapted through time, across cultures, between texts, and so on" ("Towards" 6). Tracing the development of the zombie trope will shed light on why we continue passively on our path toward the edge of a precipice despite a growing number of cli-fi narratives in the twenty-first century.

Indeed, Stephen Prince notices "a huge boom in the production of apocalypse movies that occurred after the millennium" (31). The cycle of apocalypse films has changed over time, of course, offering new generic hybrids (for instance, blending disaster with satire and comedy), more diverse racial and gender representation, as well as more experimental narrative techniques (the main hero dies at the end) (*Downsizing* (2017) or *Don't Look Up* (2021)). Ideally, this variety would reach a wider audience and raise awareness about the ongoing environmental crisis. However, having analyzed numerous apocalyptic films in his book *Ecological Film Theory and Psychoanalysis: Surviving the Environmental Apocalypse in Cinema* (2021), Geal reaches the pessimistic conclusion that Hollywood action films are unable to foster transformative behaviors. Even though they depict various catastrophic events, such narratives do not lead to sustainable lifestyle choices. Through their formal and narrative qualities, they "essentially suggest that the spectator would survive a similar real-world environmental disaster" (2) thanks to his or her resourcefulness, putting emphasis on individual action over collective responsibility. Viewers tend to identify with a few survivors, in particular the charismatic hero, rather than anonymous antlike human victims the camera shows in passing. Hence, they can enjoy the CGI spectacle of apocalyptic destruction, secure in the knowledge that all will be well in the end (2). Prince agrees that "More often than not, [apocalypse films] are tales of survival; they tell us how we can surmount catastrophe" (54). To that effect, some of them feature a "lifeboat," sometimes in a literal sense (Geal, *Ecological* 2), such as a huge ark in *2012* (2009) or an alien spaceship in *Knowing* (2009). In the former, a humble and resourceful writer saves his family from annihilation, and they eventually join a handful of fortunate survivors, the wealthiest and most powerful, on a monumental ark, sailing towards a bright new future. Even if the hero dies, as in *Knowing*, he still manages to save his male offspring. The entire planet may burn to dust, but humanity will continue to thrive in another galaxy far away. The spectator can rejoice in the final happy-end coda, with the image of destruction being displaced by the beauty of a foreign planet in another solar system where friendly aliens transport two human children, conveniently of the opposite sex and not related. Thus, the displeasure associated with the ecological disaster is substituted by what Stephen Heath calls "the jubilation of the final image" (514).

The zombie subgenre follows the classical action and disaster film tropes described above. Typically, the apocalypse is triggered by an unfortunate sudden event, such as the spread of a deadly virus (*28 Days Later*, *World War Z*) or a fungal infection (*The Last of Us* (2023–). Sometimes, the cause remains unclear (*Night of the Living Dead*, *Zombieland*, *Walking Dead* (2010–2022)), resulting in speculation, unease, and paranoia. Composed of various genre clichés, such as a captivating hero, a time-sensitive storyline, special effects, and intensified continuity, such productions build a sensation of speed and urgency. In *World War Z*, for instance, Brad Pitt's character, initially skeptical about his involvement, ends up saving his family and the entire planet from the zombie outbreak thanks to his acute observational skills, research abilities, and team leadership. Fast-paced zombie attacks spare the protagonists. Individual resourcefulness and courage are emphasized, as is the human need to rebuild communities or the semblance of family (*The Last of Us*, *Shaun of the Dead* (2004), *Zombieland*, *28 Days Later*). Living among enraged zombies requires quick thinking, spectacular escapes, swift planning, and prompt decision-making, leaving no room for indecisiveness or rumination.

However, the risk of climate change, as Geal points out, "is unfolding in slow motion" (*Ecological* 1), unlike a fast-paced Hollywood action film. This is because "Climate change is incremental, proceeding silently and in ways that may be unrecognized until a threshold is reached where changes become exponential and undeniable" (Prince 118). Eva Horn writes how "The present feels as if it is drifting toward a disaster, one almost impossible to anticipate, for it will consist not in a 'big bang' but in nothing more than inadvertently crossing a critical threshold or reaching a tipping point" (229). She proposes a new term, "a catastrophe without event," to describe "a new type of threat" (55) that "does not take place as an event but rather as a slow, imperceptible shifting of states, with multiple, contradictory, and unpredictable consequences" (56). Hence, Geal suggests that the only type of cinema that could represent this new kind of threat and raise awareness would have to perform one of two tasks—it would have to be either highly reflexive or belong to the movement of slow cinema (*Ecological* 243). Although he does not discuss Jarmusch's film in his analysis, *The Dead Don't Die* fulfills both of Geal's conditions, being both reflexive and slow.

According to the *Critical Dictionary of Film and Television Theory*, "reflexivity" or "self-reflexivity" "describes the process by which a film or television programme draws attention to itself, reminding the spectator of its textuality and status as a media construct.... It is the process by which texts foreground their authorship and production, acknowledging their status as representation" (Utterson 525). In 2008, I argued that metacinematic constructs had the potential to "wake us up from... zombification propagated by Hollywood narrative mode and the viewing circumstances in the cinema" because they could "encourage critical self-awareness in the spectator" by making us aware of film conventions and codes (178). Jarmusch's film is doubly reflexive, from numerous homages to Romero's oeuvre seen in its plot, mise-en-scène, and dialogue, through characters discussing zombie films, to, finally, their disclosure that they are in fact in a zombie film, which breaks the fourth wall. Romero's *The Night of the Living Dead* and other horror classics are discussed by a few younger characters who use their cineliteracy to show off and argue over minute details of the plot. Three young "hipsters" from a large city who drive into town in an old Pontiac LeMans like the one featured in the 1969 classic are looked at with envy. Horror buffs instruct others how to kill zombies, using their film knowledge as a reference. Ironically, their familiarity with zombie lore does not guarantee survival because in Jarmusch's universe zombie references boil down to empty quotes. Knowledge does not lead to understanding or appropriate action when it is used as a point of self-validation or trivia cataloguing. Jarmusch reinvigorates the zombie cycle, however, lifting it above and beyond Hollywood's prerogative to entertain, and mocking its circular postmodern referentiality.

For Jarmusch, "Genres are just frames—you can put whatever you want inside" (qtd. in Fear 2019). *The Dead Don't Die* is not his first foray into genre filmmaking or monster narratives, as he previously blended elements of the Western with a slow-cinema aesthetic in *Dead Man* (1995), while his *Only Lovers Left Alive* (2013) explored the life of culturally sophisticated centuries-old vampires Adam (Tom Hiddleston) and

Eve (Tilda Swinton), who call humans "zombies" because of their disregard for the environment. Jarmusch's vampires are endangered species, struggling to find blood that is not contaminated as a result of industrial capitalism. Toward the end of the film, they observe how people have severed their link with nature and, as a result, pollute rivers, lakes, and oceans, forgetting that eighty percent of their blood and more than fifty percent of their bodies are composed of water. *The Dead Don't Die* continues and develops this theme further, blending the zombie genre with slow cinema by giving it a novel ecological spin.

Despite George Bluestone's somewhat reductive claim that, unlike the novel, the film has only one tense (48), the film as a temporal medium has a unique capacity to manipulate our sense of duration. To create the impression of slowness, long takes and panning shots are preferred over close-ups. Instead of intensified continuity, editing tends to be minimal and the camera movement fluid, slow, or even static. Frequent repetitions of dialogue, sequences, and shots create a sense of redundancy, boredom, or uneventfulness. Characters appear emotionless and passive, choosing to "observe rather than act" (Jaffe 4), which explains why the plot may never reach narrative closure. Jarmusch is recognized as one of the masters of the movement, committed to "ordinary, unheroic characters and events" (Jaffe 16) in opposition to "over-blown action scenes" and "overly dramatic scenes" (Jarmusch, qtd. in Jaffe 16). In *The Dead Don't Die* reflexivity meets slow-film aesthetics, resulting in a provocative and ironic film, whose message is even stronger than that expressed in *Only Lovers Left Alive*.

The action takes place in the sleepy rural Centerville, where the sluggish police chief Cliff (Bill Murray) and his partner Ronnie (Adam Driver) cruise around the empty town where nothing seems to happen. The film opens with what appear to be two static shots of a local cemetery to the sound of a few discordant guitar cords. No discernible change or movement can be detected at first until a police car enters the frame. In this way, Jarmusch establishes a pensive, sombre mood and an atmosphere of stillness and quietude. The depopulated landscape is strangely eerie, suggesting that the apocalypse may already be under way without anyone paying it any attention. In the fifth minute of the film, Ronnie and Cliff remark that it is still daylight even though it is past eight p.m.—a reference to Romero's film, which starts with a similar observation minutes before the first zombie attack. Ronnie then realizes that his watch has stopped. Time becomes elusive and impossible to track by either modern technology or the earth's rotation. Clocks and conversations about time typically help orient the viewer as to the progression of the narrative and the passage of time. They create a sense of urgency and speed. Setting up deadlines by which the hero needs to achieve his/her goal raises the stakes and creates a growing sense of excitement and anticipation of the approaching resolution. Here, these time indicators serve another purpose—they alert us to the fact that the time is out of joint, but in Jarmusch's world, there is no hero to set it right.

Confused by the unprecedentedly long daylight, Cliff fancies coffee and doughnuts until Ronnie points out that it is already evening. The latter observes that "Something weird is going on" and then adds, "This isn't gonna end well." These words spoken in the sixth minute become our first clue that the film may be more serious than its light-hearted trailer indicated. The lines address what Nicole Seymour describes as a "deeply weird current moment—in which reports of immanent collapse inspire not robust environmentalist action but doomsday fatigue" (57). The words "weird" and "strange" punctuate *The Dead Don't Die*, being repeated sixteen times in total, as the characters wonder why the day seems longer, their phone signal is lost, the battery dead, their pets run away, watches stop and, in the end, the dead rise from their graves and start feeding on the living. The townsfolk wait for the night to fall and wonder if the recent disturbance in the day-night cycle could be caused by polar fracking. On the radio, the energy company tells them not to believe the so-called scientists, who are liars and alarmists, and to appreciate instead how many jobs polar fracking has created, helping the economy.

The story duration equals plot duration as there are no noticeable ellipses, unfolding over the course of twenty-four hours. The film's running time is only 1:44, thus shorter than *World War Z* (1:56), *28 Days Later I* (1:53), and *Army of the Dead* (2:28). And yet *The Dead Don't Die* provoked mixed reactions, with viewers criticizing it for being painfully slow, boring, and uneventful as its screen duration was experienced as lasting much longer than a regular-length feature. This is because, as

Prince notes, the "dilation of time" in slow cinema "produces a heightened, concentrated perception of duration in viewers, in part because it diverges from the conventional ways that time is constructed and experienced in cinema" (115). Typically, the duration in most mainstream apocalypse productions is not extended but rather condensed into an endless cycle of spectacular events, deleting the dead time in between. In *24/7: Late Capitalism and the Ends of Sleep*, Jonathan Crary examines how "billions of dollars are spent every year researching how to reduce decision-making time, how to eliminate the useless time of reflection and contemplation" (40), preventing the possibility of "daydream or of any mode of absent-minded introspection that would otherwise occur in intervals of slow and vacant time" (88). Here, by contrast, the director builds an atmosphere of uneventfulness, quiet resignation, and passivity through numerous techniques, affecting our sense of duration.

Many shots, lines of dialogue, and words are repeated, including the theme song *The Dead Don't Die* by Sturgill Simpson, which accompanies the opening credits and then plays on repeat within the diegesis as different characters listen to and comment on it. Such absurdist meta-jokes produce the effect of distancing, interruption, and discontinuity, initially creating humor based upon such rarely seen redundancies. They also serve as our only respite from the bleakness of the director's vision before the joke wears thin. When Murray's character gets irritated with the song, he exhibits a rare show of emotion; for the most part, Cliff and Ronnie remain remarkably stoic, if not apathetic and indifferent, in the face of the unusual events. Upon investigating the first zombie kill in a local diner, their body language and tone of voice do not reflect the gravity of the situation, despite the fact that Cliff claims it is the worst sight he has ever seen.

It is the younger Ronnie, however, who from the start appears most accepting and resigned to his fate. As he reveals to the disgruntled Cliff toward the end of the film, this is because he knows everyone dies in the end, as Jarmusch showed him the script. Jarmusch toys with generic paradigms through casting, teasing us with the trope of the hero's journey. Thus, even if Ronnie constantly reminds us that the story will end badly, he is played by Driver, allowing the viewer to hope that he will transform from a reluctant hero into a charismatic leader, taking charge against all odds. Still, Ronnie's attitude toward the crisis remains unchangeably reconciled. Even if he takes pleasure in chopping zombies' heads off, his actions are not the outcome of a carefully planned offensive but rather fill up the time before the inevitable end arrives. By contrast, Mandy, the third police officer, portrayed by Chloë Sevigny, stands out due to her visibly horrified, wide-eyed and loud reactions. She exhibits a visceral response to disemboweled corpses. Her behavior is the accepted norm in disaster films as she looks to her male colleagues for reassurance. She is also the only one to seek a rational explanation, desperately clinging to hope that armed forces will arrive and rescue them from the invasion. Her presence in the film serves as a stand-in for the viewer used to such generic parameters.

When the zombies strike at night, the locals and visitors alike lock themselves alone indoors and observe the unfolding action from the window of a house, gas station, motel, or police car. Jarmusch mocks another convention of the disaster genre, showing that when individual initiative takes precedence over collective action, it paradoxically reduces the characters' chances of survival. Cliff, Ronnie, and Mandy cruise around town and watch the zombie apocalypse unfold without any offer of assistance, already assuming the worst. Crary references George Herbert Mead, who in the 1920s came up with three constituents without which society would collapse—neighborliness, helpfulness, and cooperation (125). These elements are absent in *The Dead Don't Die*, implying that society is broken and not worth saving.

The motif of waiting and watching rather than acting is amplified as most characters are shown either passively listening to interrupted news broadcast on the radio or watching its fuzzy image on television screens. Writing about slow cinema, Tiago de Luca notes how the on-screen act of waiting that is then experienced collectively by spectators in the cinema matters especially today. Such a communally shared impression of time may raise awareness of our commonly shared space and time by restoring a sense of proximity (40). Emre Çağlayan also proposes that slow cinema can "make us think and feel differently and have the potential to shape our way of looking at the world" (ix) even though, as Ira Jaffe notes, such thoughts may often feel unpleasant (4), which is what Jarmusch excels at.

The confrontation between the two officers and zombies at the local cemetery closes the film. The scene does not represent humanity's fight for survival, being instead a gesture of resignation and acceptance of the predestined scenario. Cliff and Ronnie choose to die surrounded by the hordes of the living dead, realizing that there is no point in resistance. As they kill the local living dead, they awkwardly mouth apologies, depriving the scene of the usual gore, fast action, or high drama, and replacing it with a melancholy and somber note. The finale thus deprives us of the jubilation of the final image or the restoration of order and family. It teasingly plays with a possibility of office romance between Mandy and Ronnie at the beginning to then again subvert our expectations. Ronnie shows no qualms in chopping Mandy's reanimated corpse's head off even as she mumbles his name. In Jarmusch's world, there are no couples or nuclear families to be found, which not only deprives the film of heterosexual romance in its subplot but also presents the modern human condition as defined by singlehood, individuality, and loneliness.

The motif of the lifeboat is also cruelly mocked. An alien ship is summoned to the cemetery by an eccentric new undertaker Zelda (Tilda Swinton), who turns out to be extraterrestrial. As she walks towards the beam of light emanating from it, the convention dictates that she will save the heroes, joining them in the fight in the name of interplanetary kinship. Zelda, however, swiftly departs in her spaceship, leaving humanity to its own destruction. The privileged spectatorial gaze is undermined, too. To Ronnie's surprise, the film has an alien plot twist Jarmusch never shared with him. If Ronnie thought he knew the script and possessed superior knowledge to other characters, the joke is also on him. In this way, Jarmusch indicates that the viewer's masterful point of view (like Ronnie's) does not imply knowledge or understanding. The characters blindly follow the course, without a thought or a question, much like the zombies in their simple pursuit of satisfying their hunger.

As Olney observes, "the mass appeal of the living dead today is rooted not in our perception of their difference from us, chiefly, but in our recognition that on a fundamental level they are us" (10). Jarmusch's zombies are similar yet different from those in other productions. Unable to pounce or sprint, they give us more time to contemplate what animates them and what, if anything, differentiates them from us. For one, they gravitate toward things they knew and loved when they were alive, giving them a touch of individuality. They move awkwardly and trance-like as they attempt to mow the grass, play tennis, or shop for toys and sweets. They are also able to articulate one word that defines their deep desires: chardonnay, fashion, candy, Wi-Fi, or sneakers. Implying that they are reanimated by their consumerist pursuits is not a novel idea but rather a tribute to Romero's second installment in the series *Dawn of the Dead* (1978), which took place in the mall to satirize our consumerist addictions. Similarly, Danny Boyle's fast-moving enraged zombies in *28 Days Later* commented on "our predisposition to rage as twenty-first-century consumers" (Olney 66). In Jarmusch's interpretation, anger and rage give way to passivity and numbness as the defining sentiments of late capitalist societies.

Classic zombie lore consists of a nightly setting as the time of zombies' resurrection. Except for functioning as an ideal horror mise-en-scène, the reasons why the dead are reanimated at night are hardly ever discussed. Because Jarmusch's film is slow, it gives us time to contemplate the implications of their nocturnal activity. Their restlessness in the middle of the night speaks to our modern condition defined by Crary as "a principle of continuous functioning" (8). Twenty-first-century capitalism requires the dissolution of the boundaries between "sleep and waking, illumination and darkness" (20), to enable uninterrupted production and consumption. Crary finds that "A 24/7 environment has the semblance of a social world, but it is actually a non-social model of machinic performance" (9), which is demonstrated by Jarmusch when the undead rise from their graves at night and start filling up the empty streets, stores, and parks. Centerville is reanimated with a semblance of life, community, and sociability, making the town more "alive" at night than during the day. Since sleep is perceived as hindrance to constant activity, the twenty-four-hour existence is more "aligned with what is inanimate, inert, or unageing" (10). Eradicating sleep and becoming a zombie is the answer to the demands of twenty-first-century capitalism, as it dissolves "the brutal discrepancies between the temporal operation of deregulated markets and the intrinsic physical limitations of the humans required to conform to these demands" (15). Ironically, the first zombies woken from their slumber demand coffee. Jarmusch's posthuman automated creatures, "severed from the cyclical temporalities of lunar and solar movements" (62), cannot be put in rest mode, replicating the permanent flow of capital, uninterrupted operations of global markets, and constant functioning of digital appliances. In this way, as Crary concludes, 24/7 existence is "inseparable from environmental catastrophe in its declaration of permanent expenditure, of endless wastefulness for its sustenance, in its terminal disruption of the cycles and seasons on which ecological integrity depends" (10). Jarmusch adapts the classic zombie figure to address our lethargic attitude to climate crisis as we continue to appease our anxiety with mindless consumerism that fills our days and nights.

Responses to the film were mixed. Neither A. O. Scott of *The New York Times*, Owen Gleiberman of *Variety*, John Nugent of *Empire*, nor Wendy Ide of *The Guardian* picked up on the climate crisis analogy. Many reviewers focused instead on the film's inferiority to Romero's zombie prototype, analyzing its generic modifications, aesthetic qualities, and deadpan performances, similarly to Jarmusch's characters who brag about their comprehensive film knowledge but are unable to see a bigger picture. A few that recognized the director's message responded more positively, observing, for example, how it works as "an unspoken answer to our hubris, our sins, and our inability to recognize that our time is up, and that we're the ones who accelerated the timetable by being ignorant, greedy, and clueless" (Seitz). For Brian Eggert, the film depicts "humanity's continued sleepwalk through the apocalypse." Tim Grierson believes that "Jarmusch's zombies represent everything from global warming to alt-right groups" and notes how "In our modern world, zombies are just one more terror we've learned to normalize." Of course, global warming is hard to represent on screen in event-based drama. What Jarmusch succeeds in doing, however, is approximating a feeling of quiet anticipation, denialism, passivity, and watchful anxiety that, as the creator of Acolytes of Horror remarks, "matches my own real-life relationship with climate change" ("The Dead Don't Die: How Deadpan Dooms Us").

In sum, instead of generic tokens of fast action, tight deadlines, heroism, and special effects, the avant-garde director offers us a narrative and a narrative mode that more closely reflect the current climate fiasco. If disaster films have taught us to anticipate one catastrophic event, the apocalypse in *The Dead Don't Die* is strangely quiet and uneventful, without momentum or catharsis. Unlike his many predecessors, he does not explain it as a result of a mysterious pathogen or an unfortunate accident but openly blames human activity for the environmental disaster. Thus, even though there is hardly any anger expressed in the film, several reviewers have noticed that *The Dead Don't Die* is Jarmusch's angriest film to date. "I'm just sick of zombies," the director said, "Like, the real zombies that are just walking around us, not paying attention to anything, letting the end of the world happen" (qtd. in Fear). His timely addition to the rich zombie catalogue shows the ecocritical potential of adaptation, which can return to classic sources and reuse them to new critical ends. By employing two powerful cinematic tools, layered reflexivity and slow-movie aesthetic, his provocative take on the zombie apocalypse gives us enough time and space to pause and reflect. Whether such reflexive and slow films can foster change remains to be seen. While the jury is out, time is running out.

Works Cited

Bluestone, George. *Novels into Film*. Johns Hopkins, 1957.

Çağlayan, Emre. *Poetics of Slow Cinema: Nostalgia, Absurdism, Boredom*. Palgrave, 2018.

Crary, Jonathan. *24/7: Late Capitalism and the End of Sleep*. Verso, 2013.

The Dead Don't Die. Directed by Jim Jarmusch, Kill the Head, 2019.

"The Dead Don't Die: How Deadpan Dooms Us." *Acolytes of Horror*. www.youtube.com/watch?v=AaztKHi3Id0. Accessed 14 Nov. 2024.

de Luca, Tiago. "Slow Time, Visible Cinema: Duration, Experience and Spectatorship," *Cinema Journal*, vol. 56, no.1, 2016, pp. 23–42.

Dendle, Peter. "The Zombie as Barometer of Cultural Anxiety." *Monsters and the Monstrous: Myths and Metaphors of Enduring Evil*, edited by Niall Scott, Rodopi, 2007, pp. 45–57.

Eggert, Brian. "The Dead Don't Die." *Deep Focus Review*, 19 June 2019, www.deepfocusreview.com/reviews/the-dead-dont-die/. Accessed 6 Nov. 2024.

Fear, David. "'I'm Just Sick of Zombies, Man': Jim Jarmusch on 'The Dead Don't Die'." *The Rolling Stone*, 14 June 2019, www.rollingstone.com/tv-movies/tv-movie-features/jim-jarmusch-the-dead-dont-die-interview-847447/. Accessed 14 Nov. 2024.

Geal, Robert. *Ecological Film Theory and Psychoanalysis: Surviving the Environmental Apocalypse in Cinema*. Routledge, 2021.
Geal, Robert. "Towards an Ecocritical Adaptation Studies." *Adaptation*, vol. 16, no. 1, pp. 1–12, 2023, https://doi.org/10.1093/adaptation/apad001.
Grierson, Tim. "*The Dead Don't Die*: Cannes Review." *Screen Daily*, 15 May 2019, www.screendaily.com/reviews/the-dead-dont-die-cannes-review/5139365.article. Accessed 14 Nov. 2024.
Heath, Stephen. "*Jaws*, Ideology and Film Theory," *Movies and Methods*, vol. 2, edited by Bill Nichols, U of California P, 1985, pp. 509–514.
Horn, Eva. *The Future as Catastrophe. Imagining Disaster in the Modern Age*. Translated by Valentine Pakis, Columbia UP, 2018.
Jaffe, Ira. *Slow Movies: Countering the Cinema of Action*. Columbia UP, 2014.
Olney, Ian. *Zombie Cinema*. Rutgers UP, 2017.
Prince, Stephen. *Apocalypse Cinema*. Rutgers UP, 2021.
Rasmus, Agnieszka. *Filming Shakespeare, From Metatheatre to Metacinema*. Peter Lang, 2008.
Seymour, Nicole. "Toward an Irreverent Ecocriticism." *Journal of Ecocriticism*, vol. 4, no. 2, 2012, pp. 56–71.
Seitz, Matt Zoller. "The Dead Don't Die Movie Review." *Roger Ebert*, 2019, www.rogerebert.com/reviews/the-dead-dont-die-2019. Accessed 14 Nov. 2024.
Utterson, Andrew. "Reflexivity," *Critical Dictionary of Film and Television Theory*, edited by Roberta E. Pearson and Philip Simpson, Routledge, 2005, pp. 525–526.

13

Settler Ecologies and Western Adaptation: Unfitting Characters in *The Ballad of Buster Scruggs*

Susan Kollin

In their Netflix anthology *The Ballad of Buster Scruggs*, Joel and Ethan Coen adapt classic Western stories while also incorporating short pieces they wrote over the past twenty years. Their 2018 movie has been described as a "brutal salute to the Western," where the "specter of human mortality looms" over many of the segments and where there is "barely a forehead that doesn't get a bullet in it sooner or later" (Bradshaw). The film redirects stock elements of the Western, such as the epic encounter between heroes and villains, the persistence and fortitude of white settlers on the wagon trail, and the determinations and triumphs of a California gold miner. Refusing to deliver action in ways that might be expected in the genre, many of the plots appear unreal and cartoonish, highlighting a reluctance on the part of the filmmakers to conform and adapt themselves to predetermined ideas of the Western. In doing so, they help reconfigure the genre, laying bare its white settler

S. Kollin (✉)
Montana State University, Bozeman, MT, USA
e-mail: Susan.kollin@montana.edu

P. Demory (ed.), *Ecoadaptation*, Palgrave Studies in Adaptation and Visual Culture,
https://doi.org/10.1007/978-3-032-09847-4_13

fantasies in order to reveal how toxic beliefs often shape dominant understandings of regional and national identity. Satire and humor—frequent trademarks of their movies—help shift the core of these stories, recasting the meanings of the genre in order to address the underside of its myths.

In the Coens' film, normative conventions of whiteness and settler understandings of the region are dismantled in order to call forth new storytelling possibilities. Rather than replicate the triumphant stories of regeneration and renewal that often underpin the Western, the Netflix movie treats these understandings of region as destructive and damaging. Emerging in the nineteenth century as a response to radically changing conditions under industrialization and modernization, the popular Western frequently addressed nostalgic yearnings for "wild" nature or rural places that were located far from the urbanized East. The genre often unfolded through a series of dualisms between good and evil, nature and culture, and "savagery" and "civilization" (Cawelti 57–63). Over the years, the Coen brothers have developed what critic Ben Walters calls "an almost compulsive engagement" with the genre, with *Buster Scruggs* (2018) following their earlier Western films—*Raising Arizona* in 1987, *No Country for Old Men* in 2007, and *True Grit* in 2010 (32). Noting the achievements and range of their cinematic efforts, Walters argues that the Coens seem to "have long understood" how the Western can be "more quirky and capacious" than its status in recent years might suggest (32).

Rather than provide a tale about epic adventures in the heroic West, the movie offers a cast of mostly white settler characters who struggle to adapt to their environments, usually with great effort but little success. These figures encounter bleak futures in a space where there is little accountability to community or other living things, a place where "justice, heroism, or hope" have gone extinct, leaving them with "no window for prosperous futures," as critic Janne Lahti argues (82). These characters do not experience regenerative natures or an idealized West but instead grapple with survival and a sense of being out of place, where they seem to exist without purpose or meaning. While the Coens have been accused of taking things too far in their films, with at least one reviewer questioning whether the filmmakers "troll" their audiences (Kenny), critic Michael Atkinson sees meaning in what he calls the "Coen paradigm."

As he notes, it often seeks to expose hypocrisies and shallow beliefs, such as East Coast narcissism in *Burn after Reading*, "Minnesota nice" in *Fargo*, or 1960s suburban materialism in *A Serious Man*. What might seem like "cheap shots" to some audiences serve as insightful critiques of ill-advised behaviors to others (Atkinson 44). Such observations may be extended to *Buster Scruggs*, especially in its treatment of various unfitting and maladaptive characters whose misguided relationships with the more-than-human world and each other typically end in tragedy and death.

The popular Western arose in the nineteenth century, an era of slavery, dispossession, violence, and genocide as well as what Antoine Traisnel describes as the "mass displacement of wildlife and the accelerated extinction of animals in the United States" (2). Critics like Stephen Tatum note that the Western responded in diverse ways to these profound racial, political, and environmental crises. Drawing on Stuart Hall's understanding of the popular, Tatum points out that the genre was never a monolithic form, but rather a terrain of struggle involving multiple and competing understandings of identity, belonging, and place (158). Throughout its development, the Western has been adapted for various purposes and has experienced numerous transformations in the process. In *The Ballad of Buster Scruggs*, for instance, white settlers often face estrangement from the natural world, where they struggle with a mismatch between fantasies of the West and what they actually experience in the landscape.

As ecocritics tell us, stories circulate in our ecosystems where they carry influence and may intervene in processes of adaptation. While some stories may help us adapt and survive, other stories function otherwise. In calling for an environmental approach to the cinematic study of adaptation, Robert Geal argues that the field can help critics address fundamental concerns about the human place in the more-than-human world, and in doing so, may direct us to larger ecological insights. What is it about our culture, he asks, that "causes us to endanger our own survival by treating the biosphere with such contempt?" (6). Geal builds on the work of literary critic Linda Hutcheon who notes that "stories... propagate themselves when they catch on; adaptations—as both repetition and variation—are their form of replication" (qtd. in Geal 6).

Because of this, she explains, "the fittest [stories] do more than survive; they flourish" (qtd. in Geal 6).

Of course, there's no guarantee any particular story that persists about a location, or the American West in this case, will be a story we might wish to linger. Geal notes that within ecophobic culture, sometimes the fitness of a text may be seen in how it aligns well with instances of human exceptionalism or desires for mastery (7). Stephanie LeMenager addresses similar concerns about genre and narration in our moment of climate crisis. She argues that to "know oneself as embodied in a deep, evolutionary sense may bring about both a profound remembering of ecological enmeshment and an invitation to forget one's... relationship to power" (230). Some of this knowledge may entail the extinction of certain forms of storytelling or require laying to rest ways of narrating human relations with the more-than-human world that are damaging and unsustainable. As Jeffrey Jerome Cohen argues, throughout history we have witnessed how "culture-bound modes of story live for a while, flourish, catalyze new narratives, go dormant or extinct" (27).

While stories circulate in the ecosystem where they may help shape action or create meaning, stories also produce affects and emotions that likewise carry power. Throughout their anthology film, the Coens work against the emotional grain of many settler stories, which frequently "reproduce a form of American collective memory" that as Kevin Bruyneel argues "shields white Americans" from coming to terms with the nation's racist legacies of the past as well as the ongoing "crises of the present" (13). In their films, they often foreground a sense of alienation that arises in encounters with ordinary, mundane events and frequently address the experience of feeling oneself at odds with one's larger cultural, political, and physical surroundings. In *Buster Scruggs*, such alienation disrupts dominant narratives of the West, encouraging audiences to disengage from settler ways of thinking and feeling, what Bruyneel describes as "imperatives of belonging, status, security, profit, power, and safety in relationship to these lands" (5). Enabled in part by affect and emotion, these settler imperatives often live on and persist, reproducing narrow understandings of the past while normalizing structures of power in the present.

Throughout *Buster Scruggs*, the filmmakers confront normative affects by delving below the surface of acceptable emotions to make visible what is repressed by settler culture. Featuring an emotional landscape that has been regarded as disturbing to some audiences, the Coens' movie seems to offer a showdown between the extremes of normosis and psychosis. Normopathy, or the pathological desire to conform, was first described by Joyce McDougall in her book *Plea for a Measure of Abnormality*. Political psychologist Christopher Bollas extends the definition, describing normosis as "a particular drive to be normal" (22) with a normotic person as "someone who is abnormally normal" and whose "introspective capacity has rarely been used" (23). Critics have tended to focus attention on the psychopaths in the Coen brothers' films, less so the normopaths. Yet many of their movies feature characters who express pathological levels of conformity or excessive forms of normality that prove to be maladaptive and destructive.

Tracing this element in their movies, viewers might be tempted to ask how much the Coens drew from life experience growing up in a Jewish family from the suburbs of Minneapolis in developing their critiques of mainstream America. In interviews, the filmmakers have spoken about Minnesota, a place Joel Coen once described as similar to "Siberia" but with "family restaurants" (Dussere). The comment speaks to a sense of bleakness and social isolation that references more than just the climate and that may suggest an understanding of place that doesn't fit with stereotypes of a state known for its "niceness." In fact, there may be strong connections between the affective conformity of "Minnesota nice" and the maladaptive concept of normosis that circulates in the ecosystems of many of their films. Minnesota nice gained new critical attention after the death of George Floyd in 2020. A racially coded term that refers to the forced politeness of white residents across the state, Minnesota nice can be used as a way to "sidestep conversations about social issues" and at times, serves as a "form of denial that can perpetuate institutional racism" (Moen).

In *American Niceness: A Cultural History*, Carrie Tirado Bramen develops a genealogy of the concept, explaining how the sentiment of niceness "figured into a national fantasy of American exceptionalism," with qualities such as "friendliness" often emphasized as a national trait over less positive features such as "military might" or "economic prowess" (4). She builds on Bollas's notion of "violent innocence" which involves a projection that places negative behavior or emotion onto the other so that the self can remain innocent and virtuous (7). As a "national default mode" (7), American niceness is always tied up with its opposite—nastiness (3). Bramen argues that niceness "implies that Americans are fundamentally well-meaning people defined by an essential goodness. Even acts of aggression are framed as passive, reluctant, and defensive acts to protect oneself against the potential aggression of another" (8).

Such behavior appears in *Buster Scruggs*, as various white settlers in the Coens' film engage niceness as a preferred "mode of sociality," a response that starts out as a tool for survival and adaptation but eventually leads to their demise (Bramen 34). In "The Gal Who Got Rattled" sequence, based on a 1901 short story Western by Stewart Edward White, audiences encounter forms of niceness that cover over and hide aggressive settler behaviors including land grabs and racial hatreds. Often the characters' conformity and eagerness to adapt themselves to the fantasies of others leads them astray or even to their own deaths. In the sequence, a young woman named Alice (Zoe Kazan) abides by the cultural scripts given to her at home in the East and later in the West. She joins a group of white settlers who make their way across Indigenous lands, moving through these spaces with a blankness and passivity associated with normosis. The dutiful character follows orders so well that she later puts a bullet to her own head as she was directed to do earlier by her settler companions in the event that they are attacked by Native Americans on the trail.

While niceness as a normative and maladaptive affect in *Buster Scruggs* leads to death, settler greed is an emotion that threatens to bring some life forms to the brink of extinction. The sequence "All Gold Canyon" was adapted from a 1906 short story by Jack London that originally gained popularity in the context of the early twentieth-century environmental movement. The sequence features a miner in California (Tom Waits) who searches for gold in a secluded valley, leaving a path of destruction in his wake. Sylvan Goldberg notes that American fictions appearing in an earlier period during the nineteenth century addressed similar activities, contributing to what he calls an "extractive imagination" that helped justify the narrative of US expansion and that became central to the settler story of the America West (1221).

The extractive imagination also helped form the basis of what Kaitlin Reed calls the "California Story," a narrative that circulated in the ecosystem, leaving enormous destruction in its wake (27). In *Settler Cannabis: From Gold Rush to Green Rush in Indigenous Northern California*, Reed explains that the "California Story presents a narrative of nineteenth-century California history as a heroic tale of how the West was 'won'" (27). The gold rush "established the rush mentality" and helped set in motion "settler-colonial ideas about land and natural resources" that "persist and shape the present, specifically regarding natural resource extraction and management" (29). Reed notes that the event, "fueled by a desire to transform the land and resources of California into monetary wealth accumulation," resulted in one of the biggest migrations in human history as well as an "apocalyptic violence" aimed at Indigenous people (26). Miners' diaries indicate how the extractive imagination also encouraged settlers to view "the resources of California as inexhaustible," which resulted in a frenzy of harmful activity that continues to have consequences for the land and Indigenous people (28). Reed elaborates on the concept of rush: "As a noun, a rush refers to a sudden and quick movement toward something, typically by many people," she explains. "As a verb, to rush means to move with urgent haste—a rush to immediate profit without regard for the life of Native people or future generations. This rush mentality of intense and urgent desire for wealth... could not be met by gold alone; settlers continued to extract other resources, such as timber, fish, and oil" (30).

Yet other cosmologies and understandings of life and land have also existed that foreground a different relationship between humans and the more-than-human world and that focus on adaptation and interdependency. In *Braiding Sweetgrass: Indigenous Wisdom, Scientific Knowledge, and the Teachings of Plants*, Robin Wall Kimmerer (Potawatomi) writes about an "ethic of reciprocity" that was almost "cleared away along with the forests," the "beauty" of this relationship "traded away" for different "stuff," as Indigenous understandings of the land became threatened with extinction in the process of settler colonialism (200). In *Settler Cannabis*,

Reed reminds us that "Indigenous relationships to and understandings of place are rooted in kinship and reciprocity. Within this epistemological framework, land is alive and possesses its own agency," which highlights a "fundamentally different" conception from settler-colonial views of "land as inanimate property" (27–28).

In the Coens' adaptation of "All Gold Canyon," the extractive imagination undergoes revision. Here we are given an updated story of human entanglements with nonhuman agencies within a new context of environmental urgency. We are provided with a narrative that features not just human but also animal intelligence and cognition. We are presented with an adaptation that unfolds within the meaning and context of the Anthropocene, defined as the "Age of Humans," or more accurately as Justine Moore argues, the Capitalocene, defined as the "Age of Capital" (77). In "All Gold Canyon," the extractive imagination appears visually, with the holes the miner digs looking very much like graves across the landscape. In London's story, the narrator describes the process of extracting gold in precisely this manner. "Four feet, five feet, six feet, he dug his way down into the earth" (London 286). Each hole the miner digs is a hole representing death. We are digging our own graves, environmental scholars would argue, with our greed, desires for mastery, and instrumentalist uses of the natural world.

Throughout the Netflix adaptation, animal presences are also pronounced. Within the popular Western, we might expect to see horses, mules, and dogs, but perhaps not a performing bird that can count nor a scolding owl who teaches a greedy miner about reciprocity and respect. At one point, the miner in "All Gold Canyon" climbs a tree to gather eggs for his dinner. He takes more than just a few. An owl who is watching from the branch of a nearby tree darts a scowling look at him. The gold miner starts to put the eggs back in the nest. "How far can a bird count," he mutters and then begrudgingly says to the owl, "Okay, just one" (Figs. 13.1 and 13.2).

In *The Genius of Birds*, Jennifer Ackerman points to recent studies that outline the complexity of bird cognition, including their ability to count. Certain birds appear to be "born Euclideans," possessing a capacity to use "geometric clues and landmarks" as a way to "orient themselves in three-dimensional space, navigate through unknown territory, and locate

Fig. 13.1 An owl keeps watch over a greedy gold miner in the "All Gold Canyon" sequence of *The Ballad of Buster Scruggs*

Fig. 13.2 Trying to raid all the eggs from a nest, a gold miner realizes he's been watched by a nearby owl in *The Ballad of Buster Scruggs*

hidden treasures" (4). Noting that other birds seem to be "born accountants," Ackerman draws on a study showing the ability of newborn chicks to "spatially 'map' numbers from left to right, as most humans do," thus suggesting that birds share with humans "a left-to-right orientation system—a cognitive strategy that underlies our human capacity for higher mathematics" (4–5). Young birds can also "understand proportion," "choose a target from an array of objects on the basis of its original position," and "do simple arithmetic... such as addition and subtraction" (5). Instead of being disavowed or denied, animal cognition has been recognized and valued by human cultures, as Kimmerer points out. Mel Chen builds on this argument, noting that "it is only within a specific cosmology" that certain entities "lack agency," an erasure that usually occurs in "geopolitical spaces dominated by settler colonialism" (32).

In *What an Owl Knows: The New Science of the World's Most Enigmatic Birds*, Ackerman focuses on cognition, noting that owls are capable of using "math to pinpoint" their prey (xiii). These highly intelligent birds are also known for their adaptive capacities, which become particularly important in areas where human impacts have altered the land. "While urbanization is robbing good habitat from many species of owls," Ackerman notes that "Barred Owls, screech owls, and Burrowing Owls" have been "drawn to streetlights in cities and will nest nearby to skim the easy meals" (265). In just a few decades, these urban owls have "evolved variants in ninety-eight different genes important for cognitive function—genes that could well play an important role in adapting to city environments" (265). With "plasticity" in cognition and behavior, owl "populations have within them variation, adaptation, and resourcefulness that allow at least some individuals to respond flexibly to environmental change" (265).

Ackerman also explains that owls carry power as environmental actors. "As apex predators, wolves of the sky, owls serve an important ecological role, preventing rodents and other prey from crowding out other species, and maintaining balance and preserving the integrity of ecosystems. They show us the interconnectedness of living things" (267). Owls also engage in "aggressive nest defense" (267). As she points out, owls are known to "offer protection against predators for other bird species nesting nearby" and thus serve as "natural ambassadors for habitat conservation" (267). In "All Gold Canyon," the wise owl scrutinizes the California miner's behavior, perhaps in an effort to teach him the concept of the "Honorable Harvest," an Indigenous practice that tells us to "take only what is given, to use it well, to be grateful for the gift, and to reciprocate the gift," according to Kimmerer (20–21). Later in the sequence, a shift occurs when the miner adjusts his actions after seeing the owl's response to his greed, conceding with some reluctance that he is taking more than he needs.

Other gifts appear in the film too. In "The Gal Who Got Rattled" sequence, for instance, a domesticated dog from back East plays hide-and-seek with prairie dogs out West. The animals elicit laughter from Alice who takes joy in watching them play, their interactions becoming a gift in the form of a momentary break from her state of normosis. Likewise, the CGI depictions of butterflies, minnows, deer, and other animals in the canyon throughout the sequence offer a delightful cinematic take on the purple prose of Jack London's nature writing while also reminding us of the other living entities with which we share this planet. Hard-working horses, a playful dog, a counting owl, and a performing chicken make appearances in the anthology film, sometimes with humorous and sometimes with horrific effects. Watching the various sequences in the Netflix movie reminds us of the liveliness of the more-than-human world that underpins the Western and that needs to be more fully accounted for in our narratives of adaptation and survival.

Recent scholarship in ecocriticism and ecomedia studies examines adaptation as a process of telling new stories that may lead to more sustainable relations between humans and planetary life. If stories are crucial for adaptation and survival in a world of many living things, then the right stories may help us build better projects of adaptation and resilience. Environmental scholar and activist Rebecca Solnit addresses this point, arguing that the climate crisis entails on some level a crisis of storytelling. "We are often hemmed in by stories that prevent us from seeing, or believing in, or acting on the possibilities for change. Some [of these] are habits of mind" (Solnit). While the type of stories we produce and replicate remain important to consider, other critics are raising questions about the ecology of filmmaking itself. In "Is Adaptation Studies Sustainable?" Kyle Meikle suggests we need to bring the field "up to speed" by understanding the many ways that adaptation is "a de facto tactic—an everyday practice—in life during climate change" (261). Citing the work of Kyle Stine who reminds us "there is no such thing as a carbon-neutral production," Meikle draws attention to the material practices involved in making moving pictures themselves (268).

Indeed, ecomedia scholars remind us that we live in an era when nearly all aspects of our lives are shaped by media and digital technologies. In *Ecologies of the Moving Image*, Adrien Ivakhiv writes that moving images "unfold in a series of contexts... in relational ecologies that connect us all the way back to the places from which their raw materials were sourced and where they were crafted into manufactured works" (5). As he notes, "all of this takes place through a process that moves from minerals to... chemicals, plastics, and silicon chips, to shooting locations and sets, to editing suites and film distributors who deliver images on screen and desktop" (5). For Ivakhiv, addressing the process of production and consumption as an ecology not only becomes a way of recognizing the relations between organisms and their environments interacting "somewhere out 'in nature,'" but also becomes a reminder that the "nature 'out there' is always 'in here' as well" (5).

In *Film Ecology*, Susan Hayward goes on to note that new forms of storytelling do not necessarily result in more sustainable practices. She explains that while the move from celluloid to digital filmmaking brought costs down in many ways, it also transformed aspects of production, such that we often experience a "new sort of distance" and a "detachment from a sense of the real" when it comes to the digital (5). With celluloid film, huge amounts of oil byproducts and chemicals are needed to make and process the final product, along with massive amounts of water for cleaning purposes, which are often sent back to streams and rivers (11). Yet digital technologies also have an environmental footprint that needs to be addressed. Hollywood has gone largely digital, but in 2019 it was still considered the second most polluting industry in California (11). In addition to location concerns that shaped the ecologies of the earlier film industry, the environmental footprint of Hollywood's digital world includes cooling systems, the development of new software, the storage and disposal of electronic hardware, some e-waste that is shipped to poorer nations, and much more e-waste that is thrown away or not properly disposed of, where it often seeps into the ground and waterways (11). Much of this infrastructure goes unnoted and unaccounted for, Haywood argues, a point that should be of concern to those working in the environmental humanities and ecomedia studies.

In moving the field forward, Meikle makes a case for understanding how other disciplines are grappling with similar problems. He draws on work in architecture, especially the realm of sustainable design, as a useful model for ecomedia scholars (269–270). The argument could also be extended to page-to-screen adaptations and to environmental scholarship or academic publication itself, activities that likewise require the use of natural resources and thus bring into play the problem of climate hypocrisy. Speaking to this point, Tim Clark addresses the dangers of both hypocrisy and self-righteousness, particularly in a moment of climate crisis when "some irony about one's own stance and life" as inhabitants of a hypermodern world seems necessary (73).

Even streaming the series of environmental ads that the Coens directed over a decade ago sponsored by the Alliance for Climate Protection which show the idea of "clean coal" as another instance of greenwashing is an act that requires forms of energy consumption. Of course, so does the process of watching *The Ballad of Buster Scruggs* on Netflix. When we scale up to include all streaming video, this form of viewership in 2019 accounted for more than 1 percent of greenhouse gas emissions (Zeller; Marks 46). The environmental impact of streaming includes a wide range of activities such as the work of large data centers, transmission over Wi-Fi and broadband, and the use of devices which require electricity, most of which is made by a process that emits greenhouse gases (Sweney). Recognizing these realities helps "shred the ideology that digital media are immaterial" or that "'the cloud' is made of fluffy vapor," as Laura Marks points out (46, 47). These concerns belong in our conversations about developing an environmentally responsible adaptation study and are likely to remain issues that ecomedia scholars will be attending to in the years to come. Certainly, they are challenges made more apparent when we recognize what Margaret Atwood means when she explains that "it's not climate change, it's everything change" (Atwood).

References

Ackerman, Jennifer. *The Genius of Birds*. Penguin, 2017.

Ackerman, Jennifer. *What an Owl Knows: The New Science of the World's Most Enigmatic Birds*. Penguin, 2023.

Atkinson, Michael. "Schmuck of Suburbia." *Sight and Sound*, vol. 19, issue 12, December 2009, pp. 44–45.

Atwood, Margaret. "It's Not Climate Change, It's Everything Change." *Medium*. 15 July 2015. medium.com/matter/it-s-not-climate-change-it-s-everything-change-8fd9aa671804. Accessed 17 September 2023.

Bollas, Christopher. *The Christopher Bollas Reader*. Taylor and Francis Group, 2011.

Bradshaw, Peter. "The Ballad of Buster Scruggs Review: The Coens' Brutal Salute to the Western." *The Guardian*. 31 August, 2018. www.theguardian.com/film/2018/aug/31/the-ballad-of-buster-scruggs-review-coen-brothers-western. Accessed 17 September 2023.

Bramen, Carrie Tirado. *American Niceness: A Cultural History*. Harvard UP, 2017.

Bruyneel, Kevin. *Settler Memory: The Disavowal of Indigeneity and the Politics of Race in the United States*. U of North Carolina P, 2021.

Cawelti, John. "Savagery, Civilization, and the Western Hero." *Focus on the Western*, edited by John G. Nachbar, Prentice-Hill, 1974, pp. 57–63.

Chen, Mel. "Animacies." *Posthuman Glossary*, edited by Rosi Braidotti and Maria Hlavajova. Bloomsbury Press, 2018, pp. 32–34.

Clark, Timothy. *The Value of Ecocriticism*. Cambridge UP, 2019.

Cohen, Jeffrey Jerome. "Anarky." *Anthropocene Reading: Literary History in Geologic Times*, edited by Jesse Oak Taylor and Tobias Menely. Penn State UP, 2017, pp. 25–42.

Dussere, Eric. "Film Noir, Bête Blanche: Blood on the Snow in *Fargo*." *Bright Lights Film Journal*. 2 February 2022. brightlightsfilm.com/film-noir-bete-blanche-blood-on-the-snow-in-fargo/. Accessed 17 September 2023.

Geal, Robert. "Towards an Ecocritical Adaptation Studies." *Adaptation*, vol. 16, no. 1, 2023, pp. 1–12.

Goldberg, Sylvan. "Our Extractive Imagination: Natural Resources, Origin Stories, and the Nineteenth-Century US Novel." *ISLE: Interdisciplinary Studies in Literature and the Environment*, vol. 28, no. 4, 2021, pp. 1218–1236.

Hayward, Susan. *Film Ecology: Defending the Biosphere: Doughnut Economics and Film Theory and Practice*. Routledge, 2020.

Ivakhiv, Adrien. *Ecologies of the Moving Image: Cinema, Affect, Nature.* Wilfrid Laurier UP, 2013.

Kenny, Glenn. "The Ballad of Buster Scruggs." *Rogerebert.com.* 9 November 2018. www.rogerebert.com/reviews/the-ballad-of-buster-scruggs,-2018. Accessed 17 September 2023.

Kimmerer, Robin Wall. *Braiding Sweetgrass: Indigenous Wisdom, Scientific Knowledge, and the Teachings of Plants.* Milkweed Editions, 2015.

Lahti, Janne. "The Unbearable Settler West in *The Ballad of Buster Scruggs.*" *Cinematic Settlers: The Settler Colonial World in Film*, edited by Janne Lahti and Rebecca Weaver- Hightower. Routledge, 2020, pp. 77–86.

LeMenager, Stephanie. "Climate Change and the Struggle for Genre." *Anthropocene Reading: Literary History in Geologic Times*, edited by Jesse Oak Taylor and Tobias Menely. Penn State UP, 2017, pp.220–238.

London, Jack. "All Gold Canyon." *Stories of the Old West: Tales of the Mining Camp, Cavalry Troop, and Cattle Ranc*h, edited by John Seelye. U of Oklahoma P, 1994, pp. 275–291.

Marks, Laura U. "Let's Deal with the Carbon Footprint of Streaming Media." *Afterimag*e, vol. 47, no. 2, 2020, pp. 46–52.

Meikle, Kyle. "Is Adaptation Studies Sustainable?" *Adaptation,* vol. 14, no. 2, 2021, pp. 260–273.

Moen, Mike. "Ask a 'Sotan': Revisiting Minnesota Nice." *MPR News*, 21 November 2019, www.mprnews.org/story/2019/11/21/ask-a-sotan-revisiting-minnesota-nice. Accessed 17 September 2023.

Moore, Jason. *Capitalism in the Web of Life: Ecology and the Accumulation of Capital*. Verso, 2015.

Reed, Kaitlin. *Settler Cannabis: From Gold Rush to Green Rush in Indigenous Northern California.* U of Washington P, 2023.

Solnit, Rebecca. "If You Win the Popular Imagination, You Change the Game: Why We Need New Stories on Climate," *The Guardian*, 12 January 2023, www.theguardian.com/news/2023/jan/12/rebecca-solnit-climate-crisis-popular-imagination-why-we-need-new-stories. Accessed 5 October 2023.

Sweney, Mark. "Streaming's Dirty Secret: How Viewing Netflix Top 10 Creates Vast Quantities of CO2." *The Guardian*, 29 October 2021 www.theguardian.com/tv-and-radio/2021/oct/29/streamings-dirty-secret-how-viewing-netflix-top-10-creates-vast-quantity-of-co2. Accessed 20 November 2024.

Tatum, Stephen. "The Problem of the 'Popular' in the New Western History." *The New Western History: The Territory Ahead*, edited by Forrest G. Robinson. U of Arizona P, 1997, pp.153–190.

Traisnel, Antoine. *Capture: American Pursuits and the Making of a New Animal Condition.* U of Minnesota P, 2020.

Walters, Ben. "Trigger Happy." *Sight and Sound,* vol. 28, issue 12, December 2018, pp. 32–35.

Zeller Jr., Tom. "The Coen Brothers Do Clean Coal." *The New York Times*, 26 February 2009, archive.nytimes.com/green.blogs.nytimes.com/2009/02/26/the-coen-brothers-do-clean-coal/. Accessed 20 November 2024.

14

Against Nature? Queering the Romcom

Thomas Leitch

What is an essay on queering the romcom doing in a collection on ecoadaptation? Any rationale for its inclusion depends on two kinds of connections: one between the genre of romantic comedy and adaptation, and another between the ecology of nature and the ecology of what is commonly called human nature. The first of these connections has already been established in genre theory by Rick Altman, whose view of every Hollywood genre not as an objective category but as "the temporary by-product of an ongoing *process*" of "genrefication" (54, 62) strongly implies that individual adaptations are for the same reasons incidental by-products of an ongoing process of adaptation. For Altman, every genre is itself an adaptation, or more precisely an indefinitely extended series of adaptations, of models whose privileged status is always historically contingent and therefore subject to change. The second connection has long been urged by countless American lawmakers, jurists, and lobbyists, and by the Catholic Church,

T. Leitch (✉)
University of Delaware, Newark, DE, USA
e-mail: tleitch@udel.edu

P. Demory (ed.), *Ecoadaptation*, Palgrave Studies in Adaptation and Visual Culture,
https://doi.org/10.1007/978-3-032-09847-4_14

most memorably in Pope Benedict XVI's 2008 Christmas address, which "declared that if tropical forests deserve our 'protection,' then 'the human being' (defined as 'man' and 'woman') deserves it no less: 'We need something like human ecology, meant in the right way'" (qtd. in Morton 273).

This summary comes courtesy of ecological scholar Timothy Morton, who is particularly interested in Benedict's last four words, which reveal just how tendentious his definition of human ecology is. Nature and genre have long been understood as timeless universals, a status often claimed for ecology as well. But it is precisely an ecological perspective that decenters both nature and genre, revealing them as human constructions that seek to impose order on ceaselessly generative processes that resist it. Contemporary genre theory no longer treats specific genres like horror films or ritual tragedy as transcendental entities but sees them as constantly updated strategies for organizing sociocultural problems into material for mass entertainment. Morton's argument that "ideologies of Nature are founded on inside-outside structures that resemble the boundaries heterosexism polices" (274) suggests more specifically that the binaries that underlie both Nature and the romantic comedy are rooted in heterosexist practices and beliefs that seem universal only to those who wish they were universal. His reading of Charles Darwin, for whom "life-forms are made of other life-forms" and "life-forms are liquid" (Morton 275), makes Darwin sound remarkably like Rick Altman: biological species, each of them apparently stable and immutable, are by-products of the ceaseless process of evolution. Borrowing the term from Eve Kosofsky Sedgwick, Morton contends: "Queer ecology requires a vocabulary envisioning this liquid life. I propose that life-forms constitute a *mesh*, a nontotalizable, open-ended concatenation of interrelations that blur and confound boundaries at practically any level" (Morton 275; see Sedgwick 8).

Although Morton's critique focuses on holistic and reductive views of Nature and ecology, he could just as easily have been talking about popular film genres. Hollywood has historically defined both binary gender identities and heterosexual romance as sacrosanct norms whose validity depends on their foundation in Nature and human nature. One recent mutation of this putative connection stands out: the argument that since Americans have been marrying later in life and less often even as their reported levels of happiness have been declining, the solution to their unhappiness is to get married. As Rebecca Traister has noted, "policymakers have routinely imposed marriage—as if it were a smooth, indistinct entity—as a cure for the inequity, dissatisfaction, and loneliness that plague this nation."

The leading route to Hollywood marriage goes through the romantic comedy. Not coincidentally, the romcom may be the genre most consistently dependent on binaries: not the heroes and villains of superhero movies or war movies, but the heterosexual men and women whom the genre ferries from initial discord to sexual consummation through a binary structural process that Virginia Wright Wexman has aptly called creating the couple. These binaries have already come under more critical scrutiny in romcoms that queer the genre's conservative formula by expanding it, decentering it, deconstructing it, or adapting it to less conservative ends. "In recent years," Wexman observed in 1993, "acceptable outlets for sexuality have broadened so that the heterosexual couple is no longer seen as the sole norm in American society" (183). Pamela Demory has explored the ways this broadening has made it possible for "recent independent romantic comedies [to] play with, imitate, mock, subvert, critique—and *queer*—the conventional love narratives that are so endemic in Western cultural production." Despite the more frequent appearance in recent years of queer romcoms from *But I'm a Cheerleader* (1999) to *Fire Island* (2022), however, the genre has largely continued to reinscribe the heteronormative binaries that give the personalities and relationships of queer characters prominence over their sexual preferences and practices, repeatedly marking female lovers, for example, as butch and femme.

Instead of surveying the history or the current state of queer romcoms, I propose to reimagine adaptations of the romantic comedy that replace the binaries typical of its genre and its view of human nature with a mesh that not only accommodates but celebrates nonbinary unions: those between trans characters, characters who resist cis labels, gender-fluid characters, and polyamorous characters seeking nonbinary relationships. Despite its flirtations with several of these patterns, the Hollywood romcom has so far preferred to recuperate them, and the lovers who embody them, into familiar binary patterns.

This binarism is a feature, not a bug, of romantic comedy. After all, the target audience for the romcom is sharply divided over the question of whether nonbinary identifications should be accepted, and even after states stopped the individual bans of movies that had flourished in the earliest years of the cinema, Hollywood studios have historically been reluctant to make movies that would not play well in Florida or Texas. And of course romantic comedies depend on a foundational structure that remains stubbornly binary in its demands on the characters' desires and outcomes, tailoring them to the desires and expectations of their audience, who want every romcom to end by blessing an ideal couple who have been at odds with each other throughout the story. Even more important, romcoms assume that the erotic unions they celebrate are natural and inevitable, so that all the challenges to these unions posed by the second acts of these romcoms are presented as so many complications that can be guaranteed to yield, and yield to, binary unions as ritualistically predictable as they are natural. But as it becomes increasingly obvious that the audience for romcoms is actually a collection of audiences whose erotic expectations and desires may be quite different from the ones that traditional romcoms take for granted, it becomes increasingly clear that the romcom is less a reflection of amatory practices and fantasies than a compulsory prescription for them, and it makes sense to consider how a less prescriptive generation of romcoms might address their audiences.

Queer romcoms would face a formidable list of challenges. These only begin with the erototropic binaries baked into the characters despite their sexual orientations and practices, binaries that are abundantly on display in pioneering queer films like *The Boys in the Band* (1970). In addition, there are the structural binaries that drive the narrative, the target audience's demand for an ending that broadly implies that the couple will live happily ever after, and the fact that nonbinary audiences have so far been too limited to make alternative romcoms widely popular. Instead, they are budgeted, produced, and marketed as niche films. Lesbian films in particular are widely criticized for having been produced by and for heterosexual men. In other words, the genre remains fundamentally conservative.

This is hardly surprising, since genre as such is fundamentally conservative. Film genres typically work structurally by devising plots that license the audience's taboo fantasies in the middle before domesticating and recuperating them in the end. As Robert Warshow notes in his seminal essay "The Gangster as Tragic Hero," this is exactly what the gangster film, "expressing that part of the American psyche which rejects the qualities and demands of modern life" (130), does until the final reel, when both the gangsters and their audiences are asked to pay for the transgressions they have enacted or empathetically imagined. At least since *A Midsummer Night's Dream* (1596), romantic comedies have similarly allowed their characters and audiences to play with the conventions of romance in their middles by constructing meshes only to confirm their foundational binaries in their endings. Queer romcoms might well follow Alfred Hitchcock's thrillers, which resolve their plots without denaturing the nightmare power of their middles, by reversing this polarity, delivering resolutions to their plots while still leaving the audience nostalgic for the meshes of their magical, dreamlike middles. If this sounds improbable, there is a much more obvious way to end romcoms without closing off the possibilities they open: by leaning into the other term in romantic comedy and making laughter at least as important as romance.

Queer romcoms that lean into the second term in romantic comedy, itself a rarely remarked binary, by generating actual comedy add a distinctly new dimension to forebears like the feature films *Brokeback Mountain* (2005), *Moonlight* (2016), *Carol* (2015), and *Call Me by Your Name* (2017), along with the Hulu series *Love, Victor* (2020–2022) and the Netflix series *Heartstopper* (2022–), most of which have been cast as melodramas or tragedies rather than romcoms like *Love, Simon* (2018) and *Red, White, & Royal Blue* (2023). *Love, Simon* ("Rated PG-13 for thematic elements, sexual references, language and teen partying") in particular provides a comprehensive catalogue of the obstacles many gay teens in its audience have seen to coming out themselves and a textbook of strategies for dealing with those obstacles. The logistics of these strategies are as fanciful as you would expect from the film's status as popular entertainment, but both obstacles and strategies reveal a clear pattern of takeaways.

The film's opening voiceover ("I'm just like you. I have a totally, perfectly normal life. Except that I have one huge-ass secret"), which seems originally to be addressed directly to the presumably "normal" audience, actually turns out to be a proleptic echo of the first post Simon Spier (Nick Robinson) sends to the pseudonymous "Blue," actually Bram (Keiynan Lonsdale), the classmate who's come out online—a move that conflates the two "just like you" audiences, straight (or undetermined) and gay. Tellingly, the film ends not with Simon and Bram's top-of-the-Ferris-wheel kiss but with the new signed social media posts Simon's friend Leah points out that imitate him, establishing him as a force for good in the larger community; with an extended shot of him in the kitchen with his family before he heads to school, establishing their unremarked acceptance of him as perfectly normal and natural; and with a sequence in which he picks up his friends Nick and Leah, his formerly aspiring girlfriend Abby, and Bram, establishing Simon as a prime mover in a new community defined by the mesh of its internal conflicts. In the process the film, which differs from most other romcoms in placing less emphasis on the formation of the couple than on one party's struggles for acceptance by himself, friends, family, and cohort, suggests that it might be rewarding to adapt other romcoms that way too, shaping and reading them as stories that create couples as ways of exploring individuals' search for acceptance within their larger communities, beginning with a community of two.

But I'm a Cheerleader similarly plays with and against the binary thinking that roots both heterosexual romance and realistic casting and performances in assumptions about Nature assumed in the film's opening grace over dinner, in which Megan's father prays for the acceptance of roles that are "natural and healthy and sacred." When Mary S. Brown (Kathy Moriarty), the dragon of True Directions, asks Megan (Natasha Lyonne) whether she's ever had "unnatural" reactions to moments when she saw women in tight skirts or applying lipstick, she replies, "I don't think that's unnatural." Minutes later, when she realizes that her reactions can't be natural because they aren't universally shared, she cries in anguish, "I'm a homosexual!" True Directions' antigay strictures (e.g., the sign "Silly faggot, dicks are for chicks") are comically undermined by casting Mink Stole as Megan's mother and RuPaul as her instructor in a move that echoes Wexman's premise that the changing "Hollywood conventions of acting relate to changing styles of courtship and marriage," in particular her argument that American cinema's increasing openness to nonbinary sexualities is linked to performance styles that share "an anti-realist bias: rather than present a complete world that the audience shares vicariously by means of a 'fourth wall' convention, they approach the film as a performance in which the audience does not witness a representation but participates in a presentation" (ix, 284).

Rejecting gender binaries more wholeheartedly, as Janet Hardy and Dossie Easton advocate in *The Ethical Slut* (1997), reveals that "there are as many ways to be sexual as there are to be human, and all of them are valid" (293)—a stirring call to sexual freedom that implicitly confronts the romcom's structural imperative of definitive closure. One encouraging sign that sexual license and structural closure, two impulses constituting a tricky binary themselves, may be reconciled is the romcom's persistent ability to adapt over the years despite its inherent conservatism, a tendency attested by historical developments in the genre. After years of writing and staging anticouple comedies from *The Taming of the Shrew* (1591) to *All's Well That Ends Well* (ca. 1598), Shakespeare produces *Much Ado About Nothing* (1599) and *Twelfth Night* (1601), a pair of romcoms as exemplary for the genre as they are atypical of the playwright. Much later, Hollywood supplements romcoms revolving around virginal heroines right out of Jane Austen with the more sexually experienced heroines of Stanley Cavell's comedies of remarriage in the 1930s and 1940s, whose "plot is not to get the central pair together, but to get them *back* together, together *again*" (2). Despite Brian Henderson's claim that the romcom is so sex-proof that Burt Reynolds's line to Jill Clayburgh, "How come we never fucked?" from *Semi-Tough* (1977), marks the end of the genre, whose existence "depends on the suppression of this question" (324), a later generation of romcoms demonstrate that sexual consummation need not have an exclusively terminal valence. Films like *When Harry Met Sally…* (1989), *Pretty Woman* (1990), and *Friends with Benefits* (2011) present sexual relationships that are merely a prologue, and sometimes an obstacle, to romance. When Jeffrey (Steven Weber), the sex-hungry hero of *Jeffrey* (1995), refuses the romantic overtures of Steven (Michael T. Weiss), who's been HIV-positive for five years, the threat to the couple isn't a third party but the specter of AIDS, and it becomes increasingly clear that Jeffrey will end by accepting the risk of AIDS because to deny it by hiding from Steven would be hiding from life itself.

If the genre has adapted to changing audiences and social realities so many times before, there's no reason to assume it cannot adapt again by branding queer relationships just as natural as binary, heterosexual relationships. Encouraged by some notable precursors, this essay concludes by examining in greater detail four particular kinds of changes it has already hinted at.

1. Most obvious is the greater changes in binary gender *identity* encouraged by the remarkable profusion of cross-dressing roles across performative media, beginning with the necessary and playful cross-dressing casting of roles like Cherubino in Mozart's *The Marriage of Figaro* (1786) and Octavian in Richard Strauss's *Der Rosenkavalier* (1911). Cinema abounds in cross-dressing moments like Marlene Dietrich's onstage performance in *Morocco* (1930) and romcoms like *Tootsie* and *Victor/Victoria* (1982). In *Ma vie en rose* (1997), the behavior of seven-year-old Ludo (Georges Du Fresne), who dresses as a girl and insists that he's a girl, is considered shockingly abnormal until his family moves to a new town to follow his father's inferior new job. A closing sequence in which Ludo changes costumes under pressure from his boyish neighbor Christine Delvigne (Raphaelle Santini), who wants to swap her fairy-princess party dress for his male outfit, reveals that his new, economically downscale community thinks his behavior is no big deal, a perspective his parents are happy to adopt. Ludo's behavior never changes; nor does anyone in either community except for his parents change their minds; the only thing that changes is his geographical location.

Cross-dressing, which has a long Hollywood history as a staple of slapstick comedy, from Charlie Chaplin's *A Woman* (1915) to the many film and television adaptations of the 1892 stage play *Charley's Aunt*, represents a mesh of much broader gender fluidity or nonbinary identifications in a broad range of silent films starring women documented by

Rebecca Bell-Metereau in *Hollywood Androgyny* (1993) and Laura Horak in *Girls Will Be Boys* (2016). A widely overlooked example of male cross-dressing is the short films of Laurel and Hardy, in which Stan and Ollie often play female versions of themselves, as when Stan masquerades as Mrs. Hardy in *That's My Wife* (1929) and when Stan and Ollie are married to each other's sisters in *Twice Two* (1933), their wives played by Laurel and Hardy but their voices dubbed by Carol Tevis and May Wallace. What is most revelatory about these subversive performances is how closely the duo's signature mannerisms fit their female characters, from Hardy's mincing finger gestures to Laurel's silently anguished weeping. Even when Laurel and Hardy are not in drag, or playing against the female versions of themselves in *Twice Two*, they have often been claimed as precursors of the Gay Pride movement because they are still not exactly paragons of masculinity. As Mark Simpson acutely observes,

> Laurel and Hardy are not "gay". But they are clearly not "straight" either. Attempts by gays to claim them as "the ultimate gay couple" almost miss the point. Laurel and Hardy's dalliance with perverse signifiers—their "queerness"—is actually a measure of their gender nonconformity as much as, if not more than, a sign of sexual deviation. Their refusal/inability to perform heterosexuality and play the role of "men" is what defines them.

Simpson is not talking just about the films in which Laurel and Hardy appear in drag; he is talking about the durable personas they maintained over a twenty-five-year career: consistently male in the social roles they assumed, but much less determinate in their sexual behavior, which has led different audiences to define them as androgynous, presexual, or nonsexual, all terms that highlight the comical incongruity between the performers' personas and the gendered socioeconomic demands on their characters.

This pattern is surprisingly consistent for other male teams of film comedians, from the Three Stooges to Abbott and Costello. Although attractive women often feature as eye candy in their movies, members of these teams never take a serious amatory interest in any of them, treating each of them instead either as a sexual temptation or as just one of the boys. The biggest exception to this rule is the Marx Brothers, but even this is a very limited exception. Groucho is paired in seven films with Margaret Dumont, who projects even less sexuality than him, and after Chico, Harpo, and Groucho all pay suit to Thelma Todd in *Horse Feathers* (1932), the film's final shot shows her apparently marrying them all, sending up the binary resolution audiences might have expected.

2. More radical challenges to social assumptions about binary *gender stability* have been mounted by precursor films about characters changing their sexual identities, from *Glen or Glenda* (1953) to *The Danish Girl* (2015). Neither of these is a romantic comedy, but it is no great stretch to call the Pixar feature *WALL•E* (2008) a futuristic gender-fluid romcom, since both WALL•E and EVE, apparently gendered by their names, voices, colors, and physiques, reveal qualities that increasingly complicate this binary, echoing the male qualities that emerge in the figure Carol Clover has influentially dubbed the Final Girl of horror films, whom Clover describes as "boyish, in a word. Just as the killer is not fully masculine, she is not fully feminine" (40).

Glen or Glenda (1953) dramatizes its target audience's likely reactions to the representation of cross-dressing by setting the film's visuals against its soundtrack. The staid voiceover in its trailer stands in stark contrast to its salacious onscreen admonitions: "It's Fantastic! It's Unbelievable! It's TRUE! Most DARING FILM of YEAR," and "Sizzling! Shocking! Unforgettable! Daring and true EXPOSÉ—of a HUSH-HUSH SUBJECT!" The opening of the film moves from a bizarre "judge ye not" title to a long take in which the deeply unsettling scientist Bela Lugosi addresses the camera about the strange story about to unfold. Shortly thereafter, the insistent voiceover that shapes the audience's perception of all the visuals says over a shot of Glen walking down the street in drag: "[we all know that] nature makes mistakes. It happens every day." The shot is followed by a (presumably imagined) reaction from a different voice: "How can a guy be normal and go and do a thing like that to hisself?" The sequence proceeds to give voice to half a dozen unseen speakers whose remarks invite the audience to see the film in several broadly contradictory ways: as a sensationalistic exposé, as a case study perused by a detached clinician, as a riveting spectacle they should refrain from judging as immoral, as a specifically progressivist/modernist development comparable to the airplane, the automobile, and other new inventions initially greeted with skepticism, and as a dream or nightmare complicated by the ambivalent fantasies of Glen (Edward D. Wood, Jr., billed as "Daniel Davis") involving seductively clad women, which he finds at once horrifying and irresistible. Remarkably, most of these voices vanish from the film's fifteen-minute epilogue, a surprisingly straightforward voiceover account, selectively and decorously illustrated, that recasts Glen's cross-dressing as a prelude to the more radical journey of Alan ("Tommy" Haines), who undergoes sex-change surgery and emerges as Ann. A single shot of Lugosi making Alan disappear and Ann appear, echoing an earlier such shot of Lugosi and Glen, caps the story without reinstalling the binaries it has so thoroughly enmeshed.

3. In addition to these challenges to binary gender identity and consistency, greater changes in binary *attraction* have been encouraged by a different set of precursors that reflect the findings of Kim Parker and Rachel Minkin, of the Pew Research Center, that "lesbian, gay or bisexual adults are far more likely than straight adults to say open marriages are acceptable (75% vs. 29%)." African comedies like *Xala* (1975) celebrate or at least assume the persistence of polygamy or polyamory. Even the United States, where polygamy is illegal, has produced films and television series celebrating nonbinary relationships from the eternal threesome of *Design for Living* (1933) to the capacious household of *Big Love* (2006–2011).

Bob & Carol & Ted & Alice (1969) is especially revealing in the deeply ambivalent attitudes it adopts and encourages toward its four leading characters' sexual experimentation. The institute that photojournalist Bob Sanders (Robert Culp) and his wife Carol (Natalie Wood) visit in its opening sequence presents itself as a more natural alternative to the California culture it rejects. Even as Bob and Carol come to accept the institute's emphasis on expressing feelings over thoughts, the tight closeup of their most intimate mutual confessions tracks back uncomfortably as they're joined by more and more members of the group. This sequence sets up what will turn into a persistent but equally binary conflict involving which set of mores the characters and the audience will find more natural. When Bob and Carol first meet Ted (Elliott Gould) and Alice Henderson (Dyan Cannon) in a restaurant, their fervent attempts to get their friends to say how they really feel about Carol's outfit and Bob's hair are complicated by cut-ins of other diners looking askance at them, applying a more "natural" standard for personal decorum the film never decisively rejects. Despite its notorious poster showing the four leads in bed together, the film's subject is not polyamory but the sexual permissiveness fueled by the obsessive determination to be emotionally honest. Its climactic sequence is merely the *reducto ad absurdum* in the matter-of-fact progress Bob implies with his announcement: "First we'll have an orgy, and then we'll go hear Tony Bennett." This sequence, and the film as a whole, ends with less evidence of sexual consummation (the four wordlessly enter an elevator and proceed downstairs to lose themselves in a crowd) than of the men's persistent awkwardness (Bob looks disconcertingly into the camera), casting serious doubt on the critique of monogamy the film has apparently been advancing.

Casting a net more broadly reveals many representations in cinema and television of relational meshes that survive even though they are not based on mutual, reciprocal, and binary attraction. Although this may seem like a return to the bad old days of arranged marriages that treated women as chattel, the motif has been put to more comically constructive uses in screwball comedies like *Twentieth Century* (1934) and *The Great McGinty* (1940) and in TV sitcoms whose weekly resolutions are more properly defined as resets, like the one indicated by William Demarest's resigned verdict, "There we go again," the last line of dialogue as Dan McGinty (Brian Donlevy) and the Boss (Akim Tamiroff) square off in yet another fistfight in the closing scene of *The Great McGinty*. Nonbinary romcoms might also adapt the lessons of disability cinema, since films like *My Left Foot* (1989) and *CODA* (2021), like *Ma vie en rose*, seek not to cure their heroes but rather to bend social norms toward accepting them.

4. Finally, greater changes in binary *structure* are prophesied by still other precursors. The wistful ending of *Annie Hall* (1975) and the inconclusive ending of *Manhattan* (1979) led many audiences to exclude them from their lists of romcoms long before Woody Allen's own alleged sexual history made him one of the most prominent casualties of cancel culture. Even more radical is the subversiveness of the beginning and especially

the ending of *The Palm Beach Story* (1942), as Geraldine (Claudette Colbert) and Tom Jeffers (Joel McCrea), the husband and wife who have finally reconciled after all their adventures, suddenly tell millionaire J. D. Hackensacker III (Rudy Vallee) and the Princess Centimillia (Mary Astor), the crestfallen brother and sister who've pursued them in vain, that yes, in fact, they *do* have identical twins who aren't currently seeing anyone, leading to a surrealistic final tableau of seven lovers ready to take their vows—three couples, two of them wildly improbable, along with the Princess's discarded but persistent suitor Toto (Sig Arno)—as the camera cranes out from this mesh to reveal two florid intertitles that had also graced the film's surrealistic opening credit sequence. After all, if apocalyptic calamity can serve as the basis for comedy in *Dr. Strangelove* (1964) and *Don't Look Up* (2021), surely gender rebellion, gender confusion, gender fluidity, and the refusal of resolution can do so just as well (Fig. 14.1).

Romcoms routinely entertain the possibility of alternative gender identities and erotic relationships in their middle sections, as in the archetypal romcom *A Midsummer Night's Dream*, in which Titania falls in love with Nick Bottom after he's given the head of an ass. (Their relationship, which isn't gender-fluid but interspecies, violates what many audiences would presumably consider a greater taboo.) It is only the endings of *A Midsummer Night's Dream* and other romcoms that are recuperative and conservative, like that of *Kissing Jessica Stein* (2001), in which the eponymous heroine abandons her first-time lesbian romance and settles down with her infinitely more boring male counterpart. The most transgressive feature of *Kissing Jessica Stein* is that it encourages audiences to want and expect her lesbian relationship to triumph. So one final way to queer romcoms might be to rethink the relationship between their middles and their endings by projecting the playful world of their magical, meshlike middles indefinitely into the future. In order to escape the binaries demanding that they resolve the play with gendered roles that marks movies like *Tootsie* (1982) and *Mrs. Doubtfire* (1993), romcoms might extend this play indefinitely, or close by suggesting some way outside the frame that it could be extended. In other words, romcoms might use their endings to continue to upend the resolutions

Fig. 14.1 As the reunited couple, along with two just-formed couples and an unpaired plus-one, prepare to tie the knot in *The Palm Beach Story* (1942), the camera cranes out to suggest that their happy ending is as uncertain as it is implausible

they seem to have been headed for in order to challenge the audience's assumptions about love and romance instead of affirming them.

Inspired by all the ways these precursors have already adapted the apparently conservative formula of the romcom, it is possible to imagine further adaptations with new endings that create new kinds of couples and noncouples, including endings that might reaffirm values even more conservative than heterosexual romance: showing a cross-dresser like the eight-year-old hero of *Bruno* (2000) who finds lasting happiness in cross-dressing; creating a sad ending precipitated by incompatibilities other than those of gender identification; moving more generally from gender-queer romance to gender-fluid romance; suggesting, as the BDSM office romance *Secretary* (2002) does, that play with gender roles does not end with sexual consummation; replacing the either/or structure of traditional romcoms with the both/and logic of *Big Love*; or enlisting the mesh of lower-case nature on the side of personal and shared tropisms rather than institutionally mandated social norms.

There are many ways interested parties might adapt new takes on old formulas that would recuperate the mesh of nonbinary romcoms as natural under the aegis of a "[q]ueer ecology" that "show[ed] how beings exist precisely because they are nothing but relationality" (Morton 277). They could deconstruct the binary conflict between the realistic endings romcoms hint at and the wish-fulfilling surprise endings they implicitly promise. They could follow *A Midsummer Night's Dream* in re-examining or decentering the trope of awakening from enchantment to love. They could follow *Design for Living* in re-examining or decentering the enduring desire for naughty sexual experimentation. They could follow *Kissing Jessica Stein* in re-examining or decentering our assumptions about romantic love as a resolution and endpoint that is both binary and natural, positioning it instead as a series of desires or relations that are socially constructed, not transcendent or even perhaps natural. Finally, they could re-examine or decenter their audiences' desire for Happily Ever After, or recast it, as *Love, Simon* does, as Happy For Now, since, as *Some Like It Hot* (1959) reminds us in its famous last line, "Nobody's perfect." Creators and audiences might well consider that aphorism not as a throwaway punch line to *Some Like It Hot* but as the foundation for a new kind of romcom, one that would end, like *The*

Palm Beach Story, not with the words superimposed on its final shot, "And so they lived happily ever after," but with the words that succeed that familiar tag: "Or did they?".

References

Altman, Rick. *Film/Genre*. British Film Institute, 2000.

Bell-Metereau, Rebecca. *Hollywood Androgyny*. Second edition. Columbia UP, 1993.

Bob & Carol & Ted & Alice. Directed by Paul Mazursky, performances by Robert Culp, Natalie Wood, Columbia, 1969.

But I'm a Cheerleader. Directed by Jamie Babbit, performances by Natasha Lyonne, Clea DuVall, Lionsgate, 1999.

Cavell, Stanley. *Pursuits of Happiness: The Hollywood Comedy of Remarriage*. Harvard UP, 1981.

Clover, Carol J. *Men, Women, and Chain Saws: Gender in the Modern Horror Film*. Princeton UP, 1992.

Demory, Pamela. "Queer Romance in the Movies." PCA/ACA conference paper, 12 Apr. 2012.

Glen or Glenda. Directed by Edward B. Wood, Jr., performances by Edward D. Wood, Jr., Bela Lugosi, Screen Classics, 1953.

The Great McGinty. Directed by Preston Sturges, performances by Brian Donlevy, Muriel Angelus, Paramount, 1940.

Hardy, Janet W., and Dossie Easton. *The Ethical Slut: A Practical Guide to Polyamory, Open Relationships and Other Freedoms in Sex and Love*. Third edition. Ten Speed Press, 2017.

Henderson, Brian. "Romantic Comedy Today: Semi-Tough or Impossible?" *Film Studies Reader*, edited by Barry Keith Grant, U of Texas Press, 1986, pp. 309–28.

Horak, Laura. *Girls Will Be Boys: Cross-Dressed Women, Lesbians, and American Cinema, 1908–1934*. Rutgers UP, 2016.

Jeffrey. Directed by Christopher Ashley, performances by Steven Weber, Michael T. Weiss, Orion, 1995.

Kissing Jessica Stein. Directed by Charles Herman-Wurmfeld, performances by Jennifer Westfeldt, Heather Juergensen, Fox Searchlight, 2001.

Love, Simon. Directed by Greg Berlanti, performances by Nick Robinson, Jennifer Garner, Twentieth Century Fox, 2018.

Ma vie en rose. Directed by Alain Berliner, performances by Georges Du Fresne, Michèle Laroque, Sony Pictures Classics, 1997.

Morton, Timothy. "Guest Column: Queer Ecology." *PMLA*, vol. 125, no. 2, March 2010, pp. 273–82.

The Palm Beach Story. Directed by Preston Sturges, performances by Claudette Colbert, Joel McCrea, Paramount, 1942.

Parker, Kim, and Rachel Minkin. "3. Views of Divorce and Open Marriages." *Pew Research Center*, 14 Sept. 2023, www.pewresearch.org/social-trends/2023/09/14/views-of-divorce-and-open-marriages. Accessed 19 July 2025.

Sedgwick, Eve Kosofsky. "Queer and Now." *Tendencies*, Duke UP, 1993, pp. 1–22.

Simpson, Mark. "Funny-Peculiar Men: Laurel & Hardy." marksimpson.com, 19 Feb. 2021, www.marksimpson.com/2021/02/19/funny-peculiar-men-laurel-hardy/. Accessed 19 July 2025.

Traister, Rebecca. "The Return of the Marriage Plot." *The Cut*, 22 Sept. 2023, www.thecut.com/article/why-is-everyone-so-eager-for-men-and-women-to-get-married.html?utm_source=pocket-newtab-en-us. Accessed 19 July 2025.

Twice Two. Directed by James Parrott, performances by Stan Laurel, Oliver Hardy, MGM, 1933.

Warshow, Robert. "The Gangster as Tragic Hero." 1948. *The Immediate Experience: Movies, Comics, Theatre and Other Aspects of Popular Culture*, Atheneum, 1975, pp. 127–34.

WALL•E. Directed by Andrew Stanton, performances by Ben Burtt, Elissa Knight. Pixar, 2008.

Wexman, Virginia Wright. *Creating the Couple: Love, Marriage, and Hollywood Performance*. Princeton UP, 1993.

Part V

Adapting Nature

How have elements of the natural world itself been adapted, represented, replicated in various art forms—narrative and non-narrative?

The chapters in this section all move away from traditional ideas of text-to-text or text-to-film analysis. They focus on adaptations of natural elements in the world. In all of these chapters, elements of the natural world are copied/imitated/adapted in another medium—birds to machines and to film; plants and animals to painting; island flora and fauna to video game—and then that "text" is readapted back in yet a different medium—Hitchcock's birds to algorithm and machine; painting to environmental art; video game to an actual walk in the world. Together these chapters blur the boundaries between what is artificial and what is natural, and challenge us to rethink our conceptions of adaptation and of our place in nature.

15

Biocinemimicry in *The Birds*

Christina Parker-Flynn

Black avian shapes rapidly flap and flutter across a white screen, appearing and disappearing in somewhat metrical relation to the words of the film's titles, which are swiftly pecked away almost as quickly as they appear. This title sequence of Alfred Hitchcock's *The Birds* (1963) emphasizes the constructedness of meaning in the cinema; we read the "mere graphemes" of words just as we must read these stylized black shapes appearing behind the credits as birds, which resemble, according to Christopher D. Morris, "shadows on the wall of Plato's cave" (Morris 252). Like the Aristophanes play of the same name, the film may be considered an allegory of language, according to Morris, signaling the necessity of interpretation (Morris 251). Certainly, film scholars have long attempted to determine particular meanings behind the film's eponymous birds; perhaps they are punishing mankind's destructive co-optation of the natural world, whether for harnessing the power of the atom bomb or eating far too much fried chicken. Or, maybe, the birds represent nothing at all, not a concrete object or objective but instead

C. Parker-Flynn (✉)
Florida State University, Tallahassee, FL, USA

P. Demory (ed.), *Ecoadaptation*, Palgrave Studies in Adaptation and Visual Culture,
https://doi.org/10.1007/978-3-032-09847-4_15

are a mere symbol in what cognitive scientist Donald Hoffman defines as "networks of arbitrary complexity," a "bird" equal to no more than a desktop icon, hiding complex backgrounds of information or evolution; as Hoffman argues, ignoring these histories inessential to our immediate survival is, in fact, one of human's "key adaptations" (Anderson 137).

This chapter will use birds as an instructive tool for beginning to think about adaptation in humanistic studies, opening up a small inquiry into the shared ecology of bird and humankind with an end goal of assessing their pseudophylogenic interrelationship in the cinema, and specifically in Hitchcock's eco-horror, *The Birds*.

In *The Genius of Birds*, Jennifer Ackerman introduces the idea of "shared biology," which uses other organisms as model systems to understand our own brains and behavior. Though humans may like to uphold the notion that our language system, our means of communication and disseminating meaning, is unique and far surpasses anything found in the animal kingdom, two major components of human language find antecedents in the animal world, and specifically, in genes that overlap in the brain pathways of both humans and birds. In both species, the circuit for body movement was reproduced and coopted for vocal learning in what scientists call a "duplication event," which resulted in both their and our ability to imitate sounds (Ackerman 159). This might lead us to consider the countless other ways that birds prove foundational to our language and meaning systems. Seeing them as symbols of fertility and regeneration, Ancient Egyptians venerated birds, which are featured so prominently in hieroglyphic writing—which uses the ibis, or the bird of Thoth, as its first letter—that one of its early Arabic names was "script of the birds" (Quirke).

While biologists have determined a convergence in how both humans and birds are able to imitate sounds—the foundation of language, we might also study avian species as they appear invariably in our language, literature, and culture. Doves for peace, owls for wisdom, storks for fertility, swans for beauty—the list goes on, each term becoming a specific, communicative signifier, not unlike a letter in our cultural alphabet. Not only a tool for language, the study of birds' use of metatools also features prominently as an element of what Ackerman calls "the genius of birds," and she examines their development of metatool usage as it relates specifically to vision and motion capture. In particular, birds operated as a meta or instructive tool for the invention of early cinematic technologies and are therefore foundational to the language of the moving images of the cinema.

In his published treatise, *Animal Mechanism*, French physiologist Étienne-Jules Marey proclaims his intention "to analyse the rapid acts which are produced in the flight of insects and of birds" in order to "endeavour to imitate nature" so that by "seeking inspiration from her we have the best chance of solving the problems which she has solved" (4). In order to facilitate his scientific inquiries—and inspired by Edward Muybridge's zoopraxiscope experiments in animal locomotion—Marey invented a new method of analysis: the *fusil photographique*, or chronophotographic gun, able to capture twelve rapid images in one second. The rifle allowed motion from both camera and subject, therefore enhancing the sense of movement, and Marey used this instantaneous photography method to capture birds in flight, with a seagull as his first model. The reproduction of the seagull images by heliographic process produced "excellent silhouettes" ("Instantaneous"), the reproductions of which bear striking similarity to the shadowy bird etchings that open Hitchcock's film approximately seventy-five years later (Fig. 15.1).

Early film theorists ranging from Sergei Eisenstein to Sigfried Kracauer emphasized how lifelike film could be, and thus unsurprisingly the principle of life that biology seeks to investigate finds companion in film's spectacle, insomuch that one might say, as Jimana Canales does, that film "functioned as the *materia operandi* of twentieth century cell biology" (Canales 251). Though he advocated early in his career for images in collision—a combative cinema, Eisenstein later shifted "towards an interest in the protoplasmic forms of life, in the shapes and forms of things before identity is determined" (Nesbet 142). Eisenstein believed the shot to be like a "montage cell," or the embryo of the film: "Just as cells in their division form a phenomenon of another order, the organism or embryo, so, on the other side of the dialectical leap from the shot, there is montage" (37). In fact, one can see the relationship between the foundational concept of the montage sequence and the serial representation of images in Christian Pander's formative studies in 1817 of chick embryology. Pander transformed embryology into a proper

Fig. 15.1 Silhouetted image of seagull flying captured via the heliograph process from Marey's photographic gun. "Instantaneous Photography of Birds in Flight," *Nature*, 1882 (top). Screenshot from the opening title sequence of Alfred Hitchcock's *The Birds*, 1963 (bottom)

branch of scientific inquiry, with the objective to portray movements, or rhythmic foldings, in the egg (Wellman 267); indeed, his influential work in the study of chick embryology was necessarily pictorial from the start, making visible via a succession of images produced with artist and engraver Joseph Wilhelm Eduard d'Alton movements of and within the egg. Of the many hundreds of eggs examined, Pander "selected only those that lent themselves to being set in relationship to one another and lined up serially from picture to picture—quite independently of the incubation time" (283). Here, one is struck by the commonality between Pander's science as serial illustration and the temporal elision afforded by Soviet montage theory, images cut and edited together to compress, or disobey, edicts of time in favor of constructed effect. In Pander's embryological developments, as in early montage theory, "forms are in a state of continual change, the observer initially faces an undifferentiated mass of material, and as a result, development cannot be *observed*, but is *constructed*" (282). Before Marey used birds as the subject for capturing moving images via new technologies, Pander's experiments focused on the internal embryonic development of the chick as the visual subject for what Janine Wellman calls "a new observational regime."[1] A few decades later, Charles Darwin would use his own scientific observations of the beak varieties of Galapagos finches to establish his theory of evolution via adaptation.

Biomimicry uses biological adaptations and strategies employed by other living organisms to solve challenges and fabricate new products and technologies. Even before they were subject to Marey's invention to capture moving images—an important precursory technology in cinematic history—birds have been studied throughout history to measure ourselves by and to motivate our own inventions. In the early sixteenth century, Leonardo da Vinci, who believed that science and art were inextricably related, produced more than five hundred sketched designs for "flying machines" based on his observations of the anatomy and flight of birds, now largely considered the earliest examples of biomimetics.

Inspired by the work of da Vinci and others, like English engineer and aviator Sir George Cayley, the Wright Brothers began their mechanical aeronautical experimentation around 1899, at the same point in history that the first films were being made. Observing pigeons and paying specific attention to wing warping allowed the Wrights to improve control over their first flying device, a biplane kite, which they then applied to aircraft, leading them in 1903 to become the first humans in

[1] In the early twentieth century, German scientist Ludwig Graper would use stereocinematography to enhance the three-dimensionality of the visual representation of cell movement in relation to space, depth, and differentiation in his own experimental film investigations of chick embryology.

history to successfully pilot a power-controlled aircraft. In more recent aeronautical engineering, Airbus unveiled a conceptual, hybrid-electric airliner inspired by the "efficient mechanics" of birds; their "Bird of Prey" aircraft features wing and tail structures with individually controlled feather-like tips, which help with control while minimizing drag, and a smooth wing root—the point where the wing joins the fuselage—designed to imitate the graceful, aerodynamic arch of an eagle or falcon. Airbus's "Albatross One" concept even employs flapping, hinged wing tips, directly inspired by the seabird's ability to lock and unlock its shoulders during flight ("An Airbus").

These technologies raise the question of whether such inventions that mimic the natural world are ecologically valuable moments of convergence between nature, culture, and technology, or if they might, instead, represent the proverbial albatross around humanity's neck. On one hand, biomimicry can provide ecological solutions to present and future environmental problems, offering the possibility of creating design solutions for things like conserving water, reducing carbon, or developing more sustainable materials. Living organisms have evolved well-adapted structures over time through natural selection, and what biomimicry does is imitate elements of nature to solve complex human problems and develop more "natural" technologies.

Yet birds, from the raven to the albatross, have functioned equally as harbingers of death, impending doom, and even environmental degradation. Samuel Taylor Coleridge's *Rime of the Ancient Mariner* (1798) portrays the grim consequences of disregarding our ethical obligations to the natural world via the sailor's temporally unspecified voyage, during which he famously kills the albatross and is then forced to wear the seabird around his neck. An albatross around one's neck has since become an enduring metaphor for a problem that can't be solved or a psychological burden not easily escaped. Coleridge's *Rime* has also been coopted as a testament to more modern anxieties—regarding pandemics, for instance—and other environmental atrocities. In 2011, Nick Hayes published an adaptation in graphic novel form, a modern-day reimagining of the tale as a reflection on mankind's long-sustained and current environmental abuses. The eco-fable tells the internal story of a "modern mariner" who finds himself and his crew stranded on the North Pacific

Garbage Patch, meeting a cast of characters—like a deserted ghost drilling barge—along the way. The killing of the albatross in Coleridge's poem finds direct corollary not just in literary adaptations, but also in the horrors perpetrated on current ecosystems and affecting real albatross, now one of the most endangered birds of modern times. Photographer Chris Jordan's documentary, *Albatross* (2017), which begins with a quote from Coleridge and the famous Gustave Doré illustration of the albatross at the moment immediately before being pierced by the flying arrow, depicts the heartbreaking effects of the man-made environment on these seabirds. Most famous are Jordan's shots that shockingly depict the grotesque amount of plastic found inside the belly of Midway Island albatross, rendered quite visible after the decomposition of a bird's corpse exposes its internal, plastic contents. *Rime* operated as a late eighteenth-century eco-fable, but its signature albatross has since become realistically threatened by human-induced climate change, according to Beth Flint, wildlife biologist for the US Fish and Wildlife Service, who lists rising sea levels, increased storms, and temperature changes among the biggest threats to the species (Powell).

In Rachel Carson's *Silent Spring* (1962), the birds disappear; "the hens brooded, but no chicks hatched," ushering in "a spring without voices" (2). A seminal text in the environmental movement, *Silent Spring* warns of a future when the songbirds don't sing, which is the most dangerous "silence" that we may face in our environment if we don't decrease the threat of chemical biocides against birds, who in the book function as the emblem of the loss of natural wildlife and environmental balance on a larger scale. Birdsong—which Charles Darwin famously noted is the closest animal equivalent to human language—has also been long beloved by humans, who are "deeply moved at fundamental levels by the ubiquitous music of birds" (Robbins 183). From John Keats' "Ode to a Nightingale" to Tchaikovsky's *Swan Lake*, literary and aesthetic productions have oft expressed their utter enchantment with birdsong, while some have even mimicked it in direct adaptation. Beethoven's Sixth Symphony largely mimics nature sounds, for instance, naming the instruments as birds in the score: flute/nightingale, oboe/quail, clarinet/cuckoo. The loss of birdsong represents, then, a loss of our own linguistic foundations, figuratively threatening our own meaning-making systems.

One year after *Silent Spring*'s publication, *The Birds* was released in American theaters, the only Hitchcock film to include no nondiegetic music whatsoever—in other words, the birds do not sing. The soundtrack of the film is largely reliant on mechanically-simulated, experimental bird-like sounds produced by an instrument called a Trautonium. Michael Slowik traces the development of what he calls the "sparse sonic set pieces" in Hitchcock's films that culminates in *The Birds* (71); the director's notes often called for "silence," for instance, in relation to the schoolhouse attack sequence (Allen 106). Slowik performs a detailed reading of a particular sonic patterning in the film, observing that sound becomes sparsest in the film's second half while the bird attacks increase in frequency and dominance, which indicates a developing inversion of power between humans and birds. The largely electronically produced soundtrack creates a parallel inversion of spectatorial expectations, replacing a conventional score as well as mimicking and somewhat supplanting naturalistic sounds, causing an aberrational eeriness (Allen 114).

The ornithologist in the film, Mrs. Bundy, declares that she's never known birds of different species to flock together, and indeed the film portrays all types of birds, none more mysterious than the humble chicken. The Brenner family chickens "won't eat" their feed, serving as the film's primary bird-initiated forewarning and causing Mrs. Brenner to discover the fatally bird-pecked body of farmer Dan Fawcett in his own home. These chickens operate as narrative catalyst, but also as mute and invisible harbinger, since they are visually omitted from the film. In his own notebooks concerning the rather experimental and alternative sound effects for the film, Hitchcock makes the following note that suggests the chickens should be neither seen nor heard: "Fawcett's farm. Upon Mrs. Brenner's arrival we should hear a few natural farm sounds, but definitely no chicken sounds" (Aulier 520). And yet, in his personal notes on the sound effects for the last four reels of the film, Hitchcock repeatedly uses the word "brooding." Dan Aulier appends his own editorial note in the edition, proclaiming his love for the last reel's instruction for a "brooding silence," which to him "sounds so Hitchcockian" (523). Hitchcockian, we might suggest, because even though the adjective describes something as "threatening" or "menacing," its original definition dating as far back as the fifteenth century is "the action of incubating or hatching" eggs: from Richard Huloet's English-Latin dictionary (1552), "Brodyng as hennes doth to chyckens" ("Brooding").[2] After all, Hitchcock was the son of a poultry dealer.

The third film that Hitchcock adapted from a text written by Daphne du Maurier, *The Birds* solidified a quasi-mating relationship between the author and director in terms of reproducing in/as film adaptations; the film also further crystallized an association between avian imagery and the director himself. Du Maurier's "The Birds" tells a postwar story of birds attacking—Nat Hocken, a wounded war veteran who lives in a small Cornish town by the sea, first discovers that the birds are behaving strangely and then must fight increasingly to protect his home and his family. From the first page, du Maurier introduces birds as masters of navigation, "purposeful, intent; they knew where they were bound," following "a pattern of their own" (du Maurier 59). Humans have determined that birds actually do build mental maps, migratory birds having "evolved a natural suite of navigational tools," navigating by sunlight, star maps, internal clock, and even magnetic fields (Emery 44). Du Maurier's story includes all kinds of phrases that represent the birds as "purposeful" characters and operate as possible invitations to consider avian migration patterns as powerfully mechanical, the birds "driven by the same necessity of movement... as if compelled" (60). By the end of the story, in fact, nature takes over as the machine: bested by the birds, the inversion of power seemingly complete, Nat wonders "how many million years of memory were stored in those little brains, behind the stabbing beaks, the piercing eyes, now giving them this instinct to destroy mankind with all the deft precision of machines" (100).

In both the short story and the film adaptation, the birds can be said to represent terror falling from the skies, equivalent to the dropping of wartime explosives or the towering eruptions of hydrogen bombs. However, birds' usefulness in terms of human wartime conflict, either as tools or metaphors, was not invented by du Maurier or Hitchcock; homing pigeons were used as military messengers as far back as first

[2] In his introduction for the *Alfred Hitchcock Presents* episode "Arthur" (1959), Hitchcock gives his usual address with a fowl addition, as he addresses the evening's entertainment: "Good evening, ladies and gentlemen and brooders."

Fig. 15.2 Homing pigeon wearing miniature camera invented by Julius Neubronner in 1908. *Rorhof: The Pigeon Photographer*, www.pvf.fi/rorhof-en. Accessed 11 Aug. 2025

century Ancient Egypt and more recently during World War I, with some famed birds winning medals for their courageous life-saving abilities. By the beginning of the twentieth century, in fact, pigeons were taking their own pictures. CIA-developed battery-operated cameras allowed war pigeons to act as agents of Cold War espionage, gathering visual information for wartime intelligence (Fig. 15.2). Their usefulness was still, as ever, largely defined by their directionality and geographical orientation, always returning to their homebase "as the crow flies."

Hitchcock uses du Maurier's short story as a cell to develop his film, specifically centered around the concept of avian navigation. As he famously told François Truffaut, if he liked a story's "basic idea" he'd then forget about the text and instead "start to make cinema" (71). While making the film, Hitchcock insisted on an almost scientific precision in representing "the proper geography" of Bodega Bay, as if the gaze that most compelled his cinematic vision was not male, or even human, but belonged instead to the birds themselves looking down from above. His cinematic birds become powerful antagonists not just for their evocation of wartime anxieties but also quite human ones, the characters' directionlessness in the film amplified in contrast to the birds' absolute control over the cinematic space. As John Bruns suggests, the film is so caught up with the question of human navigation that it might be about cybernetics and its foundational concepts, including the relationship between ontogenic and phylogenic learning (Bruns 149). The evolutionary development of a species through a succession of forms—adaptations, in other words—forms the crux of phylogenic learning, which is perhaps the direction this chapter has been headed all along.

While we have been discussing important moments of convergent evolution between nature and culture, as well as birds and the cinema, I shift here to focus on two main pseudophylogenic, coadaptive influences on *The Birds*, which extraordinarily converge as contradictorily divergent influences on its cinematic ecosystem—one, a famous work of art, and the other a real-life event.

One of the most recognizable pieces in art history, Norwegian artist Edvard Munch's renowned painting *The Scream,* inspired production designer Robert Boyle's initial design for the visual "mood" of *The Birds*. *The Scream* is a unified work, each element reinforcing another within the painting's ecosystem—the hellish red sky, for instance, projects the terror reflected in the central character's face. Many consider the painting an expressionistic adaptation of the painter's lived experience, Munch's own journals explaining that his most famous painting was inspired by his own anxiety-ridden interaction with an environmental shock. All "at once the sky became blood red," he wrote, with "clouds like blood and tongues of fire" hanging over the blue-black fjord, causing him to feel "a great, unending scream piercing through nature" (Olson et al.). Astrophysicist Donald Olson and his research team determined that an environmental phenomenon inspired these blood red skies, and thus, the painting—the real-life eruption of Krakatoa (modern Indonesia) between 1883–1884, which produced vivid red twilights in Europe. These researchers visited Munch's exact vantage point in modern-day Oslo, as described in his journals, to confirm that he would have been looking toward the southwest, in the precise direction of the volcanic eruption.[3]

Visually indebted to Munch's painted red skies, the film also coevolved with a real-life event. In October 2019, the Santa Cruz Museum of Natural History's monthly "Collections Close-Up" highlighted a bizarre bird invasion that occurred in 1961, two years before the release of Hitchcock's film. Any quick internet scan would likely cause you to believe that the film was directly inspired by this real-life event in coastal California; however, the Museum smartly entitled this entry "Sheer Coincidence," acknowledging that Hitchcock was already at work on an adaptation of du Maurier's story of the same name when the event occurred ("Sheer Coincidence"). "Seabird Invasion Hits Coastal Homes" read the headline of the *Santa Cruz Sentinel* on Friday, August 18, 1961, the news story illustrating the bizarre, aggressive behavior of masses of sooty shearwaters ambushing people and buildings, breaking down power lines and vomiting up anchovies on front lawns in Capitola, California, in the middle of the night (Fig. 15.3).

Eight people, Wally Trabing reported, had even been bitten by the aggressive birds who, it was suggested, may have gotten lost in the fog while mass migrating to the area for feeding season (Brekke). Soon after, Alfred Hitchcock phoned the *Sentinel* to request a copy of that Friday's paper to use as research for his then current movie, *The Birds*. Two days later, the paper ran another article about how Hitchcock was using their cover story to prepare his movie, which was being filmed 140 miles north in Bodega Bay, the director "acutely interested" in this invasion so similar to his planned film's content. Hitchcock "denied having anything to do with the feathery invasion of Capitola," instead confirming the similarity as "merely a coincidence" (Brekke). A coincidence, yet still a rather remarkable concurrence of real and reel events that allows us to examine the ever-complex environmental influences external and internal to a film adaptation.

The mysterious behavior of the sooty shearwaters in 1961 may also have had an important environmental cause only recently come to light.

[3] A more recent study suggests that Munch's painting may instead have been inspired by a different environmental phenomenon—nacreous, or mother of pearl, clouds. For more information see Fikke et al.

Santa Cruz Sentinel

Serving Santa Cruz County for More Than 100 Years

15,654

SANTA CRUZ, CALIFORNIA, FRIDAY, AUGUST 18, 1961

EIGHTEEN PAGES

Seabird Invasion Hits Coastal Homes

He Got The Bird–Or Was It Vice Versa

Thousands Of Birds Floundering In Streets

Did Lights Bring Birds Ashore

Agency Okays SL Park Pact With Shaffer

University May Use Cabrillo Facilities To Open Its Doors

Fig. 15.3 The front-page news story about California seabird invasion and its playful photo caption—"He Got the Bird–Or Was It Vice Versa"—illustrating the adaptable reversibility of causality. "Seabird Invasion Hits Coastal Homes," *Santa Cruz Sentinel*, August 18, 1961

Researchers at the Scripps Institution of Oceanography now point to the developmental neurotoxin domoic acid as the possible culprit for the bizarre aggressions and subsequent deaths of so many sooty shearwaters on that day in 1961 (Costa). Using an archive of biological samples from a local marine monitoring operation, they discovered that high levels of domoic acid–producing *Pseudo-nitzschia*, a harmful algae bloom often referred to as a red tide, were consumed by zooplankton, which were, in turn, eaten by the anchovies, subsequently poisoning the birds that ate them (Brekke). Though not necessarily linked to rising ocean temperatures at that time, the algae bloom that made the birds go berserk were likely caused by an increase in carbon, negatively impacting the natural ecosystem. Scientists suggest that in the future, with rapidly warming ocean temperatures and increased greenhouse gas emissions making the oceans more acidic, toxic algae blooms will only, inevitably, occur more frequently.

Acknowledging all of these diverse adaptive influences deepens our understanding of *The Birds* as part of a complex network of coevolving bird signifiers, especially when we recognize that, inversely, the film has since imprinted itself into developmental technology in what I would like to define as an act of biocinemimicry, a circuitous association that emerges when film becomes the (pseudo) bioelement to be mimicked and adapted. Metaheuristic optimization algorithms are often designed by imitating insects, animals, or, of course, birds. The particle swarm optimization (PSO), for instance, is an algorithm inspired by the movement and foraging behaviors of individual "particles" within flocks of birds. A more recent and most fitting optimization algorithm called the HBIA, or Hitchcock bird–inspired algorithm, was influenced by the bird attack patterns portrayed in the 1963 film and meant to "capture the essence of fictional behavior of the phenomenon of birds" cinematically depicted, which to the creators, resembles what one would expect of "natural bird behavior when faced with extreme conditions" (Morais).

Whereas avian biomimicry has historically optimized mankind's chimerical and perhaps noble dream of taking flight, new drone technology has resorted to no longer humbly learning from birds but, instead, using their bodies to manufacture exceedingly lifelike aerial robots. Recently, the Netherlands based Drone Bird Co. has engineered artificial peregrine falcons and gulls, animal impostor flying cameras meant to blend in with a flock. These flapping fabrications, enhanced with extra unrealistic features, ended up starring in the Bollywood film *Uri: The Surgical Strike* (2019). When questioned if the artificiality of their AVES gulls[4] would attract attention, CEO Jan-Willem van den Eijkel suggested that someone "would really have to be an ornithologist" or equivalent to tell the difference (Ryall). Perhaps even more ghastly, engineers at New Mexico Tech have recently turned away from using artificial materials for building these types of drones, choosing instead to re-engineer the organic matter of dead, taxidermied birds into ornithopters able to more efficiently observe wildlife with minimal

[4] The company's AVES birds take their name from the Latin "aves," meaning "birds"; see Ryall.

disruption—or, to spy on people unnoticed, just like a secret camera (Orf).

Russian formalist Boris Eikenbaum saw our interaction with cinema as a type of learned behavior, or translation experience, not unlike the duplication event that occurs in both humans and birds, allowing for the species' development of language and speech. According to Eikenbaum—seemingly the earliest to use the term "cine-mimicry"—a cinematic sequence would require translation into the film's "internal speech," accomplished by the spectator's performing the complex mental labor of connecting the shots and consequently establishing the chain of film sentence(s), interpreting the complex semantics of the film's context. As an ecoadaptation, *The Birds* presents an exemplary case of what I further define here as biocinemimicry, the film both adapting the artistry of a pre-existing literary story while likewise imitating the real, biological forms and behaviors of birds. Yet *The Birds* becomes, coextensively, foundational as its own film species, a cinematic representation of birds that has been adopted as exceedingly real in its depiction of bird behaviors and/or effects, becoming in this way a postprimordial form to then be adapted from, as in the case of the HBIA algorithm. What we consider cinematic ecoadaptations aren't necessarily limited to those caught in the one-way directional act of artificially representing real-world environmental landscapes, concerns, or disasters; instead, as this chapter indicates, cinema's lifelikeness gives way to considering how film(s) may also operate like bioelements to be re-mimicked into life and its attendant technologies and cultural understandings—more lifelike than life itself, and no more bird than *The Birds*.

Works Cited

"An Airbus futuristic conceptual airliner 'takes flight' to inspire next-generation engineers." *Airbus.com*, July 19, 2019, www.airbus.com/en/newsroom/news/2019-07-an-airbus-futuristic-conceptual-airliner-takes-flight-to-inspire-next. Accessed 7 Aug. 2025.

Ackerman, Jennifer. *The Genius of Birds*. Penguin Books, 2017.

Allen, Richard. "The Sound of 'The Birds.'" *October*, vol. 146, Fall 2013, pp. 97–120.
Anderson, Erik. *Bird*. Bloomsbury Academic, 2020.
Aulier, Dan. *Hitchcock's Notebooks: An Authorized and Illustrated Look Inside the Creative Mind of Alfred Hitchcock*. Avon Books, 1999.
The Birds. Directed by Alfred Hitchcock, performances by Rod Taylor, Tippi Hedren, and Jessica Tandy, Universal Pictures, 1963.
Bjerklie, David. "How Birds Think." *Life, Birds: The World's Most Remarkable Creatures*, May 12, 2023, pp. 22–33.
Brekke, Dan. "Acid Flashback: Santa Cruz Bird Frenzy, Hitchcock and a Biological Whodunit." *KQED*, Feb 26, 2016. www.kqed.org/news/10868506/acid-flashback-santa-cruz-bird-frenzy-hitchcock-and-a-biological-whodunit. Accessed 11 Aug. 2025.
"Brooding." *OED Online*. Oxford UP, www.oed.com/view/Entry/23745. Accessed Jan. 2025.
Bruns, John. "'The Proper Geography': Hitchcock's Adaptation of Daphne du Maurier's 'The Birds.'" *Hitchcock and Adaptation: On Page and* Screen, edited by Mark Osteen, Rowman & Littlefield, 2014, pp. 145–159.
Canales, Jimana. "Dead and Alive: Micro-Cinematography Between Physics and Biology." *Configurations*, vol. 23, no. 2, Spring 2015, pp. 235–251.
Carson, Rachel *Silent Spring*. Mariner Books, 1962.
Costa, Lucio, et al. "Domoic Acid as Developmental Neurotoxin." *Neurotoxicology*, vol. 31, no. 5, May 13, 2010, pp. 409–423, pmc.ncbi.nlm.nih.gov/articles/PMC2934754/.
Du Maurier, Daphne. "The Birds." *Don't Look Now: Stories*. New York Review of Books, 2008, pp. 59–100.
Eikhenbaum, Boris. "Problems of Cine-Stylistics." *The Poetics of Cinema (Russian Poetics in Translation, Volume 9)*, edited by Richard Taylor, et al., E-book, 2013.
Eisenstein, Sergei. "The Cinematographic Principle and the Ideogram." *Film Form: Essays in Film Theory*. Harcourt, 1949, pp. 28–44.
Emery, Nathan. *Bird Brain: An Exploration of Avian Intelligence*. Princeton UP, 2016.
Fikke, Svein M., et al., "Screaming Clouds," *Weather*, vol. 72, no. 5, April 24, 2017, pp. 115–121.
Hayes, Nick. *The Rime of the Modern Mariner*. Viking, 2012.
"Instantaneous Photography of Birds in Flight." *Nature*, May 25, 1882, pp. 84–85.

Jordan, Chris, director. *Albatross*. 2017. 1 hr., 37 min. www.albatrossthefilm.com/ourstory.

Marey, Etienne-Jules. *Animal Mechanism: A Treatise on Terrestrial and Aerial Locomotion*. Appleton and Company, 1879.

Morais, Reinaldo, et al. "A novel metaheuristic inspired by Hitchcock birds' behavior for efficient optimization of large search spaces of high dimensionality." *Soft Computing*, vol. 24, no. 8, April 2020, pp. 5633–5655.

Morris, Christopher D. "Reading the Birds and *The Birds*." *Literature/Film Quarterly* vol. 28, no. 4, 2000, pp. 250–258.

Nesbet, Anne. *Savage Junctures: Sergei Eisenstein and the Shape of Thinking*. Tauris, 2007.

Olson, Donald, et al. "When the sky ran red: the story behind The Scream." *Sky & Telescope*, vol. 107, no. 2, Feb. 2004. digital.library.txst.edu/server/api/core/bitstreams/d7449206–9b95–4245–99ae-5b612c013eca/content. Accessed 13 Aug. 2025.

Orf, Darren. "Scientists Turned Dead Birds Into Zombie Drones to Spy on Humans." *Popular Mechanics*, Feb 16, 2023. www.popularmechanics.com/technology/robots/a42940200/scientists-turning-dead-birds-into-drones/. Accessed 11 Aug. 2025.

Powell, Hugh. "Facing Into the Wind: The Complicated Fate of the Laysan Albatross." *Cornell Lab: All About Birds*, 26 Sept. 2014, www.allaboutbirds.org/news/facing-into-the-wind-the-complicated-fate-of-the-laysan-albatross/. Accessed 7 Aug. 2025.

Quirke, Stephen. "The Writing of the Birds: Ancient Egyptian Hieroglyphs Before and After the Founding of Alexandria." *Electryone*, vol. 5, no. 1, 2017, pp. 32–43.

Robbins, Jim. *The Wonder of Birds*. Spiegel & Grau, 2017.

Ryall, Zach. "A Bird James Bond Might Love." *AOPA*, 20 July 2020, www.aopa.org/news-and-media/all-news/2020/july/20/a-bird-james-bond-might-love. Accessed 11 Aug. 2025.

"Sheer Coincidence." *Collections Close-Up Blog*, Santa Cruz Museum of Natural History, October 2019, www.santacruzmuseum.org/collections-october-2019-sheer-coincidence/. Accessed 1 May 2025.

Slowik, Michael. "'Not for Tourists': Hitchcock's Sparse Sonic Set Pieces." *Hitchcock Annual*, vol. 21, 2017, pp. 71–104.

Trabing, Wally. "Seabird Invasion Hits Coastal Homes." *Santa Cruz Sentinel*. Aug. 18, 1961. history.santacruzpl.org/omeka/items/show/82839#?c=0&m=0&s=0&cv=0. Accessed 7 Aug. 2025.

Truffaut, François. *Hitchcock*. Simon & Schuster, 1985.

Wierzbicki, James. "Shrieks, Flutters, and Vocal Curtains: Electronic Sound/ Electronic Music in Hitchcock's *The Birds*." *Music and the Moving Image*, vol. 1, no. 1, 17 February 2012.

16

Material Worlds: Seventeenth-Century *Vanitas* Still Life and Contemporary Environmental Art

Kim Waale

From 1977 to 1982 artist Vija Celmins created "To Fix the Image in Memory, I–XI" an artwork presenting eleven desert stones alongside meticulously painted cast bronze replicas. Museum viewers closely study these *trompe l'oeil* illusions to tease apart the adaptation from the source. "Part of the experience of exhibiting [the duplicates] together with the real stones," Celmins said, "was to create a challenge for your eyes. I wanted your eyes to open wider" (Celmins). Having collected the rocks in northern New Mexico and brought them to her studio, she remembers developing the "desire to try and put them into an art context. Sort of mocking art in a way, but also to affirm the act of making: the act of looking and making as a primal act of art" (Garrels). It took years for Celmins to perfect the illusion. Viewers undoubtedly consider the *gap* embodied in this artwork as they examine twenty-two "stones" together in a museum case: eleven original stones, none remarkable, all displaced from the desert mixed in with near perfect duplicates, eleven fake stones

K. Waale (✉)
Le Moyne College, Syracuse, NY, USA
e-mail: waalek@lemoyne.edu

P. Demory (ed.), *Ecoadaptation*, Palgrave Studies in Adaptation and Visual Culture,
https://doi.org/10.1007/978-3-032-09847-4_16

in a vitrine where they belong. "To Fix the Image in Memory, I–XI" engages the viewer to complete the work of art by attempting to distinguish the source from the adaptation, but also to question the urge to flawlessly reproduce nature. (To commercialize it and capitalize on it? To make it entertaining? To aestheticize it? To control it, own it, unwild it, to be part of it?). Art leads viewers to imagine and question; can art also inspire personal and collective change?

This essay examines seventeenth-century *vanitas* still life paintings, which flawlessly reproduced the natural objects and cultural products that abounded in the markets and, as adaptations of these paintings, contemporary artthat responds to an environment threatened by the impact of heedless human consumption. In the seventeenth century, there was an explosion of expansive global commerce concentrated in the Dutch Republic, and as a result the Dutch encountered a profusely diverse material culture. Their still life paintings became amalgams of available goods: local and imported, ordinary and extraordinary, natural and artificial objects. Elizabeth Alice Honig writes that the seventeenth century still life painting, in part a response to global trade, "establishes a space that isolates objects from their prior meanings and uses and allows for their interplay with one another" (183). This interplay, according to Honig, is a creative process that is initiated by the collector who acquires the goods and the painter who constructs the *trompe l'oeil* pictorial spaces they fill, but is eventually a collaboration involving beholders. Honig goes on to state, "We are all thereby invited to become part of a community of interpreters of material things," a practice that is integral to much contemporary environmental art (183).

Robert Geal's notion that "adaptations offer more specific and more privileged examples of how different moments/cultures relate to ecological issues" (10) provides an ecocritical lens for comparing and interpreting "material things" or texts (films, poems, works of art, etc.) and enables us to consider what he suggests is the "fundamental ecocritical question—what is it about our culture(s) which causes us to endanger our own survival by treating the biosphere with such contempt?" (6).

In his article "Twelve Fallacies in Contemporary Adaptation Theory," Thomas Leitch lays groundwork for an expansive view of adaptation. In Fallacy #7, Leitch examines the gaps between sources and their adaptations, concluding that gaps enable "inventions that elevate adaptations above servile transcriptions" (114). The gap, then, is a productive space freeing adaptors (and engaged viewers) to create culturally—and environmentally—relevant art.

This essay considers adaptation as the adjustments made to cultural products by artists due to their environments to improve their chances of survival in that environment. In considering biological adaptation and mutation as an approach for contemplating the value and efficacy of an adaptation, Leitch writes, "Instead of looking back, biological adaptation looks ever forward. Mutations survive not because they are just as good as their progenitors but because they are better at the kinds of skills their changing environments demand" ("To Adapt or to Adapt To?" 99). In particular this essay examines how certain strategies of engaging the viewer in the work of art persist and have evolved across the nearly four-hundred-year gap between the seventeenth-century Dutch Republic and contemporary Western culture. The art and artists of both cultural moments challenged and challenge viewers to adapt their lifestyles and values to existential demands. In doing this, contemporary environmental artwork re-engages the tensions in traditional Dutch *vanitas* still lifes between desire and death, awe and shame, presence and absence.

Vanitas still life paintings are sumptuous *trompe l'oeil* representations of material culture and demonstrations of exquisite technical virtuosity which, in seeming contradiction, also warn of the futility of earthly pleasure in the face of certain death. *Vanitas*—from the book of Ecclesiastes 1.2, "vanity of vanities, all is vanity" (*Harper Study Bible RSV*)—carries meaning similar to the earlier *memento mori* ("remember you must die") paintings: all is futile in the face of death. Both depict items such as extinguished candles, decaying fruits, soap bubbles, and skulls, all referencing the brevity of life, but *vanitas* typically includes more worldly symbolism (Muzdakis). This is evident in *Still Life with a Skull and a Writing Quill* (1628) by Pieter Claesz. In this intimate painting appears an overturned glass reflecting the fleeting light of windows in the room where the artist is working, a gap-toothed skull atop papers, a writing quill, a toppled inkwell, and the smoldering wick of an oil lamp—a barely visible trail of smoke indicating it was just extinguished. Claesz arranged these objects on a stone ledge so immediately in the foreground that the viewer seems just arm's length from them—the exacting textural detail adding to the sensation that we can feel surfaces, smell residual smoke, and peer into the darkened room beyond the frame of the painting. Together these items symbolize life's brevity and represent the sobriety of the prevailing Protestant culture. Metropolitan Museum curator Walter Liedtke explains that "this kind of painting reflects not only the vanity of knowledge but at the same time, the knowledge of vanity… that the owner of this picture is actually displaying his sense of awareness that the mundane world is not that important. So, it's in a way, wearing your morals as well as your knowledge on your sleeve" (Liedtke).

The growing concentration of wealth throughout the 1600s in the Dutch Republic combined with factors including Protestantism, changed the role of artists. No longer commissioned to paint religious or historical subjects, for the first time in Western history most artists made paintings on spec (Atkins 15). Artists began to develop independent reputations and careers—selling their paintings to the newly rich mercantile class. Increased demand for art led to more artists, and this led to artists differentiating and branding themselves. As soon as new goods entered the markets, artists incorporated them into their paintings and, in turn, those who owned these paintings signaled status and wealth (Atkins 47).

Willem Kalf's 1664 *Still Life with Fruit in a Wanli Bowl* is a sumptuous painting filled with luxuries that represented good taste and prosperity: delicate Chinese porcelain, a tray made from South American silver, Persian carpet, Italian marble, Mediterranean fruit, Venetian glass, and a knife with an agate handle—presumably imported from Japan (Atkins 41). Kalf could paint thin and translucent porcelain better than anyone. He used the most expensive and hard to obtain blue pigment, ultramarine, to add value to a painting that was already a luxury item. Even so, one may view this painting as a commentary on conspicuous consumption. The sensuously curling peel of imported lemon poised to fall from the marble tabletop into the viewers' space engages us directly in this picture of wealth and tempts us to indulge in the rich sensory pleasure and fruits of prosperous life. But it also tempers our enjoyment by reminding us that just as peeled fruit will rot, life will come to an end.

Willem Claesz. Heda's 1633 *Still Life with Glasses and Tobacco* is an early example of the *toebackjes* or tobacco still lifes. By the 1630s tobacco, imported from the Americas, was popular, and Heda's addition of tobacco in his paintings nods to the overlapping markets; buyers and sellers of tobacco were also buyers of paintings (Atkins 29). In this painting, we again see windows reflected in the glass wine *roemer* indicating a dimly lit room beyond the scene. A table immediately before us holds an overturned silver tazza and several tobacco pipes, one broken; the disarray is evidence of human presence, now absence, a *vanitas* reference evoking time passing. Many of these still lifes allude to a pause in action, offering viewers a moment to think, to consider the path we are on. Further references to impermanence and ephemerality in this painting include a smoldering waxed rope for lighting pipes, a peeled lemon on a silver plate teetering on the edge of the velvet-covered table, cracked nuts, and loose tobacco on a printed, folded paper. Even tobacco and its ephemeral smoke evoke the fleetingness of life. The composition also provides opportunity for viewers to consider the global trade system that brought these foreign goods into their markets and homes. The Dutch obsession with tobacco (and sugar) was built on unacknowledged human exploitation. It's estimated that the Dutch trafficked over 600,000 enslaved Africans to the Americas until abolishing the trans-Atlantic trade in 1814 (Atkins 32). Dutch still lifes look straightforward but are complicated: simultaneously embodying presence and absence, indulgence and guilt or shame, ambition and warning.

Still Life with Oysters, Cookies and Comfits by Osias Beert, c. 1610, presents two Ming dynasty porcelain bowls overflowing with confections made of Brazilian sugar. While enslaved labor is nowhere evident in this painting, it is the dark underside of this finely set table. The exotic imports that abound signify the homeowner's considerable means. One's eye wanders among dishes laden with rare and expensive goods: sugar-coated cinnamon sticks from Southeast Asia, quince paste from Portugal, almonds from Spain, currants from Greece, and oysters (likely local). Shipping raw sugar to European ports depended on the labor of enslaved African and Indigenous people. Beert spent his career immortalizing the bounty of trade networks that brought riches from around the globe, often through brutal and inequitable means, to Antwerp (Atkins 48).

Ars longa, vita brevis; art is long, life is short. These sumptuous paintings both present and preserve the ephemeral. Wafting smoke, shifting light, decaying fruit, bursting bubbles all represent the fleetingness of worldly treasures, pleasure, and ambition, but they also immortalize the same in paint, triumphing over both time and death and fixing the artists' reputations well into the future. At the heart of the *vanitas* is a paradox. Indeed, Julie Berger Hochstrasser notes, "in reality the presence of these paintings in such affluent milieux was doing the work of conspicuous consumption, *even as* their content might have been interpreted by some as undertaking to mediate against it. It is by now the fabled paradox of still life: how can luxurious paintings mediate against luxury?" (224). This paradox positions these paintings as sources for adaptations we find in contemporary eco-artwork. Like seventeenth-century Dutch *vanitas*, contemporary eco-art can address transience and fleetingness and offer viewers the opportunity to consider mortality and the folly of ambition and desire within a context of wanton, excess consumption. While seventeenth-century Dutch artists moved away from the *vanitas* tradition to focus on their commodity-filled culture, contemporary eco-artists responding to climate change and environmental degradation have reclaimed the *vanitas* strategy to challenge viewers' values and bring about cultural change. These artworks update the coded messages in *vanitas*: pay attention to how you live because your choices may lead to the deaths of others, human and nonhuman, and devastation on a global scale. *Vanitas* may remind viewers of human temporality. Likewise, the environmental artworks in this essay strive to provoke sensations of awe, loss, and even grief to address the transience of life on the planet. They also engage time, emphasizing the accelerating pace of human-caused, fossil-fueled climate change, which is transforming the planet and leading to the sixth mass extinction. Collective human impact on the environment is creating a scenario which outpaces our ability to survive it.

Sam Taylor-Johnson's *Still Life* (2001) invokes a *trompe l'oeil vanitas* through its subject, composition, and reference to transience. However, this is a 35 mm silent, color film that uses time-lapse photography (duration 3′18″) to transform a heaping pile of beautiful fruit into a rotted mass. The remorseless, unstoppable advance of time accelerates the rotting fruit. Plus, the film loops, endlessly repeating death and resurrection. There is a complication: Taylor-Johnson placed a Bic pen in the foreground, and, like a foil to death, the petrochemical pen does not change at all. In an interview, the artist explained, "Initially, when I was filming *Still Life* I had someone overseeing it and they would go in every day to check the reels of the film and they took the pen out, thinking that I had left it in by mistake. So, having filmed for about six weeks, I had to start all over again!" (Wendt).

In another Taylor-Johnson time-lapse adaptation of a seventeenth-century *vanitas*, *A Little Death* (2002), a dead hare hangs, its leg nailed to the wall, its head on a table near a perfect peach. As the body of the hare begins to decay, the animal seems to revive due to the movement of writhing maggots, reminding viewers that life stems from death. Shot over weeks, one of its most intriguing elements is the unaffected peach: the fruit, unlike the hare, remains unputrefied. Taylor-Johnson explained, "One really disturbing aspect to me was the fact that the peach in the work was I guess a genetically modified peach. In six or seven weeks it didn't even slightly rot. There was no sense of decay at all and that was quite disturbing" (Wendt).

Taylor-Johnson's two still lifes do what static painting cannot—they animate time. The films, paradoxically disgustingly gross and exceedingly beautiful, embody a sense of inevitable loss and raise questions about what does not succumb to decay and death, and why. The gap between the source and the adaptation is revelatory of human intervention and mastery. We are now able to genetically modify fruits to be unnaturally resistant to decay and produce plastics which do not degrade for hundreds of years. These films, despite their obvious reference to both *memento mori* and *vanitas* paintings, raise the specter of a fully mastered nature where there is no decay or death to bother our minds.

According to ecocritic Alexander Phillips, we perceive and understand the natural environment aesthetically. Phillips explains that environmental problems are often also aesthetic problems, citing as an example the parable that opens Rachel Carson's *Silent Spring*, in which the people in the fictional village sense there is something wrong because they no longer hear beautiful bird songs; they are surrounded by silence.

That shift in aesthetic experience leads them to deduce that the environment is not normal, something is wrong with nature. He discusses how large-scale environmental challenges that span the entire planet, like global warming and sea level rise, are difficult for humans to perceive and understand—and even believe in—because they might not yet directly impact our immediate environs. We have aesthetic/sensory experiences that alert us to local environmental problems, like a foul-smelling river, but vast problems, "the things that are not open to our perception or senses can be happening without us knowing it" ("Intro to Ecocriticism"). Ecocritical works of art make us aware of environmental problems that may be vast in scope by providing a localized aesthetic/sensory experience.

"Ice Watch"—from Studio Olafur Eliasson—is an effort to link localized aesthetic sensations to epically scaled environmental phenomena. Since 2014, Eliasson's studio has harvested and transported dozens of enormous free-floating calved icebergs from the waters of Greenland to London, Paris, and other European cities. Installed in a clock formation in urban spaces, "Ice Watch" is environmental performance art: the icebergs melt in public, and viewers interact with them as they disappear. "You can see it melting. You can hear it making noise" said Minik Rosing, Eliasson's primary scientist collaborator (Judah). Essentially a *vanitas* still life writ large, this work (like Taylor-Johnson's *Still Life* of rotting fruit) evokes the inevitability of loss. Strongly anchored to the source of the ice, the Arctic, it conjures not a frozen, forever wild Arctic of our imaginations, but a contemporary melting too-fast Arctic resulting from human-caused global warming. Watching these immense blocks melt on the pavement in front of a museum like the Tate Modern (2019) is like watching a time-lapse film of the real Arctic melt.

The concept Homo bulla (man is a bubble), alluding to the fatal frailty of human life, is omnipresent in *vanitas*, appearing as floating soap bubbles and glass orbs. This reference is embodied in *Ice Watch*, which allows viewers to hear the popping of air bubbles formed and frozen 15,000 years ago. People lie on the bergs, hug and kiss them, taste them, and tangibly sense the melting of the polar ice cap. "Ice Melt" presents hulking orphaned icebergs as vulnerable offspring of Mother Earth and implicates viewers. As an adaptation of the *vanitas*, we enjoy the incredible aesthetic experience and yet, watching the bergs melt, we feel a range of emotions associated with shirking responsibility, guilt, and helplessness—a complex of feelings that paradoxically pits our pleasure against the well-being of the planet.

In 2021, from May through November, the artist and environmental activist Maya Lin exhibited *Ghost Forest* at Madison Square Park in the heart of Manhattan, New York. *Ghost Forest* was made of forty-nine dead Atlantic white cedar trees sourced from the New Jersey Pine Barrens—the largest contiguous forest on the east coast of the United States, which is increasingly threatened by climate change–induced conditions. These cedars succumbed to excessive saltwater intrusion resulting, in part, from sea level rise. Lin explained that the trees were "about to be cleared... so, in a sense, we are borrowing these cut trees for the duration of the installation. They will then be recycled into building materials and mulch" (Lin). At a height of thirty-five to forty-six feet, these majestic but ashen and defoliated trees were installed as a grove among and in contrast with the vibrant life of the other trees in the park, relying heavily on this healthy urban nature context for impact. According to Brooke Kamin Rapaport, curator of the Madison Square Park Conservancy, "*Ghost Forest* underscores the concept of transience and fragility, and stands as a grave reminder of the consequences of inaction to the climate crisis" (Pogrebin).

Lin and MSPC designed programming alongside the exhibition, like planting one thousand trees and shrubs throughout the five boroughs of New York City to offset the carbon footprint of constructing *Ghost Forest*. Likewise, they made available educational information about solutions to climate change, such as practices to increase carbon sequestration and preserve biodiversity. In collaboration with the Cornell Lab of Ornithology, Lin composed a downloadable soundtrack featuring the calls and songs of animals, from bears to wolves, beaver and elk, which were once common in Manhattan. Lin says, "We have become so disconnected from nature that many of us do not even realize what we are losing" (Lin).

Ghost Forest creates a tangible awareness of what we are losing. In the manner of the *vanitas* still life artist building their composition of objects, Lin hand-picked each of the forty-nine trees, looking for cedars that bore scars, explaining that she wanted viewers to connect with each tree on a visceral level (Helfand). "The trees are totemic sculptures, each one with its own character," Lin explained. "Once we installed the first, it determined where the next would go; from one grew the constellation of the remaining forty-eight. I've learned over the years that making art that looks natural is extremely difficult…. It was crucial that the installation feel organic, not regulated or designed" (Yerebakan). She added that viewers will sense that these trees are dead, "as the other trees leaf in spring, summer, fall, and then back to winter, as nature is all around you and growing and living, these will get grayer and grayer and grayer" (Helfand). Just as viewers are invited into the pictorial space of the *vanitas* still life, Lin built an art space for viewers to literally immerse themselves within. Throughout the exhibit, thousands of people walked among and spent time beneath the towering dead trees, sitting, reading, and even picnicking. Bob Williams, a professional forester and one of Lin's collaborators, was thrilled that so many people might start thinking about the impact of climate change on disappearing forests because of this immersive artwork being located in a densely populated city ("N. J. 'ghost forest'"). Even though there is a sense of apocalyptic urgency associated with *Ghost Forest*, Lin emphasizes that "there is still time for all of us to make a difference" (Lin). Reinforcing the message in his *New York Times* review of *Ghost Forest*, Holland Cotter writes, "In a goes-around-comes-around universe, the karma we're producing—through competitive greed, unthinking waste and targeted malice—is killing the world around us. We're at war with the planet and everything on it, trees included. Some people simply don't know that human-sourced climate change exists. Others underestimate its gravity. Still others — a recent United States president—dismisses it as a fiction" (Cotter).

Unlike the constructed, displaced landscape of *Ghost Forest*, David Nash's *Ash Dome* is *of* the land and grows in a secret location near where Nash lives in north Wales. In 1977 Nash planted a thirty-foot circle of twenty-two ash saplings with the intention of collaborating with nature on a sculpture that would grow, evolve, and outlive himself. Having had a job planting Norwegian sitka spruce in straight rows for a forestry company, producing, he said, "bland, alien plantations taking over the hills and moorland," Nash wanted to grow a "shape of space," a dome of ash (Nash, *David Nash* 46). This is art about patience, time, and place. Since viewers do not experience *Ash Dome* in person, it necessarily takes form in viewers' imaginations—forcing us to call upon our own sensory experiences with nature to construct the sculpture. And, like *vanitas* still lifes, it invites us in.

Using hedge-making techniques to fletch the trees, Nash hoped *Ash Dome* would mature by 2000 to serve as a memorial to a decade dominated by cold war, oil crises, and nuclear turmoil. According to the artist, "The 1970s were a grim time for humanity and many were predicting the human race would not see the twenty-first century. I felt the major problem was the short-termism of all world governments and that long-term thought and action were desperately needed. The idea of 'growing a space' for the twenty-first century was a gesture of this thought" ("Mystery").

Nash patiently tends to *Ash Dome*, intervening periodically to manage the growing form, though mostly he must wait for the trees to mature and shape the dome—it's a sculpture that cannot be rushed. He writes that "the experience has been one of direct contact with the vitality of these trees" (Nash, *Forms* 66). Throughout Nash's body of work one can recognize an unwillingness to fully master his material (always wood)—he only exerts his anthropocentric influence so far and then lets nature do what it will do. In 1978 he chainsaw-carved a spherical half-ton oak boulder. After he maneuvered it into a mountain stream, it became what he called a "Wooden Boulder with continual potential" (*Forms* 52). Occasionally, over the years, he would seek it out as it journeyed toward the sea. In 2015, Nash said the last spotting of his "free-range sculpture" was in an estuary in north Wales, and that he suspected it was now out to sea. "It is where it is," he said (Nash, interview).

In the 1990s, ash dieback, a fungal disease, arrived to Europe and eventually the UK via ash trees imported from East Asia. As a result, millions of the UK's ash trees are dead or dying, and *Ash Dome* is doomed. Nash explained, "The concept for this project was that the form should emerge from the natural energies of its environment. I felt I had to respect that basic tenet and regard the fungus as a natural force—and just hope the ash dome trees would be immune. But they are not. Since

2019 the fungus was present and now seven of the trees have died. Most of the others are badly infected by it." He went on to say, "Instead of just watching it die—which is very painful—the decision was made to start a new dome around the old one, this time with oak trees" ("Mystery"). A sculpture meant to stand as an environmentally integrated life force for the twenty-first century is instead succumbing because of globalization, reckless trade, and eco-systemic vulnerabilities. A fungal disease that eerily calls to mind the recent global pandemic has made *Ash Dome* a *memento mori* still life.

Controlled and measured, Mark Dion's *Neukom Vivarium* (2006), is an artwork constructed around a sixty-foot long dead and decaying hemlock tree that fell in an old-growth forest in the northwest United States. Transported to Seattle and enclosed in a custom-designed, climate-controlled art space greenhouse, this relocated "nurse log" in its fine-tuned biosphere is a host for hundreds of organisms, all nurtured by the decaying carcass. In its "Sleeping Beauty coffin," it's an ecosystem on expensive and high-tech life support. It's precious, it's precarious, and much life depends on it. Like the tensions inherent in *vanitas* between death and desire and awe and shame, *Neukom Vivarium* is about aestheticized tension and cognitive dissonance. Dion is foregrounding nature's cycles and processes, but unlike the trees of *Ash Dome* and *Ghost Forest,* the nurse log and its offspring are protected from invasive fungi, rogue temperature change, desertification, sea level rise, etc. We can visit this sculptural monument to biodiversity, but we cannot live in this protective biosphere. This *vanitas* still life turns the tables on the genre. The desirable worldly riches of *Neukom Vivarium* are the harbingers of death in *vanitas.* Rot and decay in Dion's *Vivarium* literally mean life.

We expect nature to persist—or at least we used to. As early as 2006 a new "stone" called plastiglomerate has appeared. Plastic, an anthropogenically derived material that first emerged in the 1950s, is now both abundant and persistent in the environment, especially accumulating in Earth's waters and along shores. A scientific publication titled "An anthropogenic marker horizon in the future rock record," explains that plastiglomerates, first found on remote Kamilo Beach in Hawaii, are a conglomerate of melted plastic, beach sand, basaltic lava fragments, corals and organic debris that most likely formed from human made campfires." The article explains, "it is conceivable that the global extent of plastic debris could lead to similar deposits where lava flows, forest fires, and extreme temperatures occur" (Corcoran). According to one of the scientists who discovered plastiglomerates, "All around the world where there's trash being openly burned in mass quantities, you can imagine there are even larger melted plastic deposits where plastiglomerate could form" (Lewis).

In 2019 in response to the unabating presence of plastic trash in the oceans, Canadian artist Kelly Jazvac exhibited plastiglomerates in Milan's Triennale as ready-made artworks which mark the Anthropocene. Finding them beautiful and horrific at the same time, Jazvac says, by showing them "as artworks and artefacts, I think they will prompt viewers to ask important questions about all the conditions and processes that led to their existence" (Yalcinkaya). As noted earlier, Vija Celmins' "To Fix the Image in Memory, I-XI" engages the viewer to decipher how a source (desert rocks) can be distinguished from the adaptation (painted replicas). The plastiglomerates selected and displayed by Jazvac engage viewers in a similar way, but they raise questions: Are we looking at nature adapting itself? Are we viewing an innocent illusion—or an ominous reality?

As the seventeenth century progressed, Dutch painters became more enmeshed in the thriving marketplace, building their reputations and amassing collectors, letting the *vanitas* content of their paintings lapse. The paintings continued to be tantalizingly lovely, but in a world that could "be produced, controlled, procured and exchanged," they ceased to foreground questions about mortality (Atkins 189). Four hundred years later, swimming in plastic on an overheating planet, artists are urgently pushing *vanitas* content: consider how you live because you will die. However, it's not by bellying up to the seventeenth-century Dutch still life table to contemplate one's own mortality amidst images of decaying fruit, skulls, and tipped-over tazzas. Instead, contemporary environmental artists create tangible sensory/aesthetic experiences to get us to open our eyes wider. Contemporary eco-art has viewers hugging icebergs, walking in dead forests, and examining plastic rocks, alerting us to serious ecological challenges beyond ourselves and urging us to *adapt* our collective habits, attitudes, actions, and relationships to better the chances that we (and our nonhuman cohabitants) will survive and, crucially, preempt what is threatening to be environmental collapse on a grand scale. By re-engaging the tensions in traditional Dutch vanitas still lifes between desire and death, awe and shame, presence and absence, contemporary environmental artwork leads us to imagine and question what it is to be human on this planet, provoking viewers towards personal and collective change.

References

Atkins, Christopher D. M., et al. *Dutch Art in a Global Age*. MFA Publications, Museum of Fine Arts, Boston, 2023.

Celmins, Vija. *To Fix the Image in Memory*. 1977–82. Museum of Modern Art, www.moma.org/collection/works/100210. Accessed Sept. 2024.

Corcoran, Patricia L., et al. "An anthropogenic marker horizon in the future rock record." *GSA Today*, vol. 24, no. 6, June 2014, https://doi.org/10.1130/GSAT-G198A.1.

Cotter, Holland. "In Maya Lin's 'Ghost Forest,' the Trees Are Talking Back." *New York Times*, July 1, 2021.

Garrels, Gary, et al., editors. *Vija Celmins: To Fix the Image in Memory*. San Francisco Museum of Modern Art, Yale UP, 2018.

Geal, Robert. "Towards an Ecocritical Adaptation Studies." *Adaptation*, vol. 16, no. 1, March 2023, pp. 1–12, https://doi.org/10.1093/adaptation/apad001.

Harper Study Bible, Revised Standard Version. Edited by Harold Lindsell, Harper and Row, 1962.

Helfand, Zach. "Maya Lin's Ghost Forest in the Shadow of Shake Shack." *The New Yorker*, 10 May 2021, www.newyorker.com/magazine/2021/05/17/maya-lins-ghost-forest-in-the-shadow-of-shake-shack. Accessed Jan. 2024.

Hochstrasser, Julie Berger. "IMAG(in)ing prosperity painting and material culture in the 17th-century Dutch household." *Netherlands Yearbook for History of Art / Nederlands Kunsthistorisch Jaarboek*, vol. 51, no. 1, 2000, pp. 195–236, https://doi.org/10.1163/22145966-90000664.4.

Honig, Elizabeth Alice. "Making Sense of Things: On the Motives of Dutch Still Life." *RES: Anthropology and Aesthetics*, Autumn 1998, no. 34, pp. 166–183.

"Introduction to Ecocriticism with Alexander Phillips." *The Trans-Atlanticist*, July 2018, season 1, episode 5, thetrans-atlanticist.podigee.io/s1e9-ecocriticism-with-alexander-phillips. Accessed Dec. 2024.

Judah, Hettie. "Was Olafur Eliasson Bringing 30 Icebergs to London a Sustainability Own Goal?" *Frieze*, 14 Dec. 2018, www.frieze.com/article/was-olafur-eliasson-bringing-30-icebergs-london-sustainability-own-goal. Accessed Aug. 2023.

Leitch, Thomas. "Twelve Fallacies in Contemporary Adaption Theory." *Film and Literature: An Introduction and Reader*, edited by Timothy Corrigan, 2nd ed., Routledge, 2012, pp. 104–121.

Leitch, Thomas. "To Adapt or to Adapt to? Consequences of Approaching Adaptation

Intransitively," *Studia Filmoznawcze,* vol. 30, no. 91–103.

Lewis, Danny. "Are We Living in the Plastic Age?" *Smithsonian Magazine*, 22 Jan. 2016, www.smithsonianmag.com/smart-news/are-we-living-plastic-age-180957817/. Accessed 20 July 2025.

Liedtke, Walter. "#5205 Still Life with a Skull and a Writing Quill." *The Metropolitan Museum of Art*, www.metmuseum.org/art/collection/search/435904. Accessed Jan. 2024.

Lin, Maya, et al. "Maya Lin Builds a Final Monument to the Climate." *Frieze*, April 28, 2021, no. 219, www.frieze.com/article/maya-lin-ghost-forest-2021. Accessed Jan. 2024.

Muzdakis, Madeleine. "Vanitas: Paintings by the Dutch Old Masters Inspired by Life and Death." *My Modern Met*, 11 Feb. 2022, mymodernmet.com/vanitas-dutch-paintings/. Accessed Jan. 2024.

"Mystery of Art Masterpiece That Drifted along River for 37 Years, Then Vanished." *North Wales Live*, 27 June 2021, www.dailypost.co.uk/news/north-wales-news/mystery-art-masterpiece-drifted-along-20903822. Accessed Aug. 2023.

Nash, David. *David Nash*. Abrams, New York, 2007.

Nash, David. *Forms into Time*. ARTMEDIA PRESS, London, 2001.

Nash, David. Personal interview. November 2016.

"N. J. 'ghost forest' becomes part of art installation in New York City." *YouTube*, uploaded by NJ.com, May 7, 2021, www.youtube.com/watch?v=jJSEEATKF2U. Accessed Jan. 2024.

Pogrebin, Robin. "Maya Lin to Conjure Dying Trees to Make a Point." *The New York Times*, 11 Nov. 2019, www.nytimes.com/2019/11/11/arts/design/maya-lin-madison-square-park.html. Accessed 1 Feb. 2024.

Wendt, Selene. "Breaking the Medium of Painting Down." *Art Pulse Magazine,* April 2011, artpulsemagazine.com/breaking-the-medium-of-painting-down. Accessed Jan. 2024.

Yalcinkaya, Gunseli. "Kelly Jazvac Presents 'Beautiful and Horrific' Plastiglomerate Objects at Milan Triennale." *Dezeen,* 29 Apr. 2021, www.dezeen.com/2019/04/21/kelly-jazvac-plastiglomerate-milan-triennale/. Accessed Sept. 2024.

Yerebakan, Osman Can. "Maya Lin on Planting a Ghost Forst in Manhattan." *Art Forum,* 22 June 2021, www.artforum.com/columns/maya-lin-on-planting-a-ghost-forest-in-manhattan-250154/. Accessed Jan. 2024.

17

The Nature of Experience and the Experience of Nature in *Proteus: Artifact Edition*

John Sanders

Spring: Introduction

> In every walk with Nature one receives far more than he seeks.
> —John Muir, *Steep Trails*

When it was initially released via digital download in 2013, Ed Key and David Kanaga's *Proteus* was unlike any other game at the time. The player begins out at sea, their virtual eyes opening to reveal the shores of a colorful, blockily rendered island in the distance. As they move toward it, the sounds of a computerized springtime rise to meet them: the modulated whistles and warbles of unseen digital birds, the melodic sound of waveform wind in the background, and soft chiming tones representing glistening dew on the blocky flowers or perhaps the square blossoms that float off the pixelated pink cherry trees. This soundscape is altered

J. Sanders (✉)
Appalachian State University, Boone, NC, USA
e-mail: sandersjw@appstate.edu

P. Demory (ed.), *Ecoadaptation*, Palgrave Studies in Adaptation and Visual Culture,
https://doi.org/10.1007/978-3-032-09847-4_17

by occasional encounters with strangely familiar digital fauna, like flat-headed frogs that produce bounding arpeggios as they leap away from the player or horn-faced worms that honk as they peek out of the ground. Though the game does have an ending and technically a way to progress towards it—each night, a portal appears in a circle of standing stones somewhere on the island that transports the player to the next season—*Proteus* offers few of the structural accoutrement we might expect from standard digital games. There are no power-ups, no puzzles, no plot, not even a way to interact with the environment except for walking through it at a leisurely pace and admiring its sights and sounds (or, occasionally, sitting down by pressing the spacebar). It is for this reason that *Proteus* and games like it earned the generic marker of "walking simulator," an originally derisive term which attempted to point to the folly of adapting the experience of an aimless stroll through nature into the form of a digital game.[1]

Although much ink has been spilled regarding how *Proteus* and games like it complicate the definitional boundaries of games and play,[2] what has been less discussed is the way that *Proteus* operates as an ecoadaptation: that is, the ways in which it shapes and conceptualizes the relationship between its virtual world, the real-world referents on which it is based, and the player who wanders between the two. Granted, there are legitimate reasons why *Proteus* is not a prominent text within the discourses of ecocriticism or adaptation studies. In the first place, digital games have traditionally been seen as an "indoorsy" medium, not least because the hardware and peripherals needed to create and play them are best positioned near an electrical outlet. While few would go so far as to blame digital games alone for drawing our attention away from the natural world—after all, the same critique could be leveled at the compelling storyworlds of film, television, or novels—the immersive audiovisual stimulation provided by artificial worlds like *Proteus* does seem tailor-made to keep players at their computer screens rather than

[1] While some developers and fans have reclaimed the term "walking simulator" in the years since *Proteus's* released, designer Ed Key has been quoted as calling *Proteus* "more of a wandering and being surprised game" (O'Connor 2014).

[2] For more on "walking simulators," see Melissa Kagen's *Wandering Games* (2022) or Sonia Fizek's *Playing at a Distance: Borderlands of Video Game Aesthetic* (2022).

exploring the outdoors. In addition to not necessarily resonating with the "eco" label, *Proteus* as described above does not seem to be an adaptation either. Aside from a sharing a title with a shape-shifting Greek demigod (a connection that proves to be incidental) and including a John Muir quote in its official trailer (re-created in the epigraph for this chapter), *Proteus* does not seem to have any explicit textual or environmental "source texts" with which it may be compared. In fact, because the gameworld is procedurally generated every time someone starts a new game, it is incredibly unlikely that the player will see the same island topography or arrangement of prefabricated game elements twice, precluding the possibility that *Proteus* could accurately represent any particular landscape at all (fictional or otherwise).

However, while the original digital iteration of the game may not easily lend itself to the label of "ecoadaptation," that is only because its sources were not yet made public at the time of its release. Three years after the original game went up for sale on electronic storefronts, the designers released a physical version of the game which included an audio CD, a photographic history of the game's development, a "Field Guide" to the fictional island (written in the style of an eighteenth-century naturalist), and a mysterious deck of cards labeled "The Protean Map." Taken together, the paratextual objects within the aptly named *Proteus: Artifact Edition* (2016) reveal *Proteus* to be a palimpsestuous adaptation of the creator's experience of the natural world, grounded in specific real-world locations and inspired by the tradition of nature walking. Aside from drawing players' attention to the world outside of the game, these objects—particularly "The Protean Map"—grant them the tools to re-enchant real-world landscapes with the same sense of wonder that inspired *Proteus* in the first place, essentially completing the adaptational and ecocritical loop initiated by the original game. Tracing the path of this loop involves exploring, first, the way *Proteus* establishes an ecocritical relationship to space and, second, how the *Artifact Edition* extends this relationship beyond the boundaries of the screen. In doing so, I aim to show that *Proteus: Artifact Edition* presents an experiential, aesthetic practice that any work of ecocritical adaptation could benefit from, one which intentionally blurs the boundary between artificial and natural spaces in order to promote mindful and meaningful engagement with both.

Summer: *Proteus* as Ecogame

Despite their reputation as promoters of what Richard Louv has called "nature deficit disorder" (36), multiple scholars have argued that when compared to other media, digital games—and by extension digital game adaptations—are uniquely positioned to get us thinking about the environments we inhabit. When comparing the affordances offered by verbal, visual, and interactive media in her work *A Theory of Adaptation*, Linda Hutcheon makes the claim that "what is often most significant for videogames is the adapted heterocosm, the spectacular world of digital animation the player enters," made all the more viscerally real by "the visual and audio effects" accompanying the player's exploration of it (51). Whether it is the hallowed halls of the titular school of magic in *Hogwarts: Legacy* (Portkey Games, 2022) or the idyllic footpaths and burrows in *Tales of the Shire* (Wētā Workshop, 2025), traversable representations of familiar settings offer a chance for designers and players to explore those elements of a storyworld that lie just off-screen or just beyond the page. Games that adapt historical or contemporary real-world environments—such as the landmark-laden historical cities of the *Assassin's Creed* series (Ubisoft, 2007–2023) or the impressively rendered pond of *Walden, a game* (USC Game Innovation Lab, 2017)—benefit from a similar attention to detail, evoking wonder in players who have never seen these places and nostalgia for those who have been there before. Whether created from fragments of imagined descriptions, well-framed televisual set pieces, or blueprints and photographs of real-world locations, the art of creating spatially coherent and audio-visually immersive synthetic worlds involves the practice of what Edward Castronova calls "selective fidelity," balancing player expectations of how an environment ought to look and feel with the needs of compelling gameplay (88). When this balance is achieved, according to Hutcheon, digital games can provide "an 'intensity of engagement' unrivaled in most other media" (51), an immersive experience which allows players to "feel physically present in the mediated environment, rather than in our real world" (135).

Putting aside for a moment the ecocritically troubling implications of this latter phrase (to be discussed further in the next section), it is important to note that gameworlds do not only alter the player's relationship to a particular piece of source material through the audiovisual representation of adapted places: they also use encoded rules to change the user's relationship to space itself. As Alenda Y. Chang claims in *Playing Nature*, a major part of games' capacity for environmental representation arises from their "knack for drawing attention to the ways in which we traverse and occupy space," regardless of the visual fidelity these spaces might have to their real or imagined sources (21). The condensed Manhattan of *Marvel's Spiderman* (Insomniac Games, 2018), for example, is not designed to be gawked at from the ground level; instead, it is designed for players to climb, jump, and swing across using the unique abilities of the game's superpowered protagonist. Every piece of architecture within the game is outfitted with invisible "anchor points" for Spiderman's webbing, encouraging players to plot routes using tall landmarks in order to traverse the rooftops in the most effective or kinetically interesting ways (Massongill). The inclusion of lower-profile districts like Queens in the game's sequels necessitated new gliding mechanics in order to keep the sense of effective traversal over areas with fewer tall buildings from which players may swing (Gratton). Beyond mere traversal, games also shift the way players think of particular spaces based on the way they are able to interact with or use them. The efficient UI and top-down perspective of a strategy game like *Dune: Spice Wars* (Shiro Games, 2023) encourages the player to visually break up the vast desert of Arrakis into discrete conquerable regions prioritized by exploitable resources (much like the imperial despots of Herbert's series do), while the survival mechanics of a game like *Dune: Awakening* (Funcom, 2025) require the player to scan the environment for sources of water and shade as they traverse these sands on foot. Far beyond their ability to immerse the player with interactive audiovisual spectacle, then, the ecocritical and adaptational potential for digital gameworlds lies in the ways they encourage us to inhabit particular relationships to space through their rules, topographies, and limitation on player action.

Though there are myriad ways to approach creating an ecologically responsible relationship to space in a digital game—scholars Laura op de Beke, Joost Raessens, and Stefan Werning include everything from strategy simulations like *SimEarth* (Maxis 1990) to first-person shooters like *Battlefield 2042* (DICE 2021) in their 2024 collection *Ecogames* (10)—the particular spatial strategies of *Proteus* can best be explored through the more condensed, four-part definition of an "environmentally oriented work" presented by Lawrence Buell in *The Environmental Imagination*. According to Buell, an environmentally oriented work is one in which:

> 1. The nonhuman environment is present not merely as a framing device but as a presence that begins to suggest that human history is implicated in natural history.
> 2. The human interest is not understood to be the only legitimate interest.
> 3. Human accountability to the environment is part of the text's ethical orientation.
> 4. Some sense of the environment as a process rather than as a constant or given is at least implicit in the text. (7–8)

By the time the player has transitioned from spring into summer, *Proteus* touches on every one of these criteria through the way it generates a navigable interactive space. For one, the nonhuman environment is far from mere background or a framing device in *Proteus*; with no quantitative tasks to complete, heroic role to fulfill, and no way of exploiting the environment for personal gain, the player's human agency is marginalized in favor of centering the environment itself. This environment is not devoid of the markers of the human, either, as many of the landmarks that reappear across every island—a lonely cabin, a ruined tower, a circle of totems, and (of course) the circle of standing stones that allow players to transition between seasons—evoke the history of an absent humanity that was once nestled between the hills and trees. The flora and fauna of the island (most abundant and varied in the summer) embody the second of Buell's criteria, as they make their own way across the island and seem to have a mind of their own. Coveys of partridges, scampering squirrels, even the oddly burbling mushrooms all run or hide themselves when the player approaches, contributing to the sense that (as designer Ed Key put it in a 2011 interview) "there's wildlife you can't quite grasp, there's a wild world and you can't go and pet it" (Yang). As I will explore further in this chapter's final section, the game's ethical orientation and the sense of "environment as process" both arise from the fact that, despite needing to wait until nightfall for a portal to appear in each season, it is the player who ultimately triggers the transition from one season to the next, irrevocably changing this environment as they progress through the game. In short, *Proteus* could be said to present a heterocosmic ecosystem, a living other world that can only be approached in a nonexploitative, slow-paced manner and which gives players ample chances to contemplate its wonders through playful interaction.

Autumn: *Proteus*: *Artifact Edition* as Ecoadaptation

While *Proteus's* reverent relationship to and vibrant representation of nature marks it as an environmentally oriented work within the bounds of its virtual world, it is actually this very vibrance and wonder-inducing reverence that reveals its artificiality. One of the most awe-inspiring and revealing moments of the game comes during a night in autumn, a season that fittingly proves to be a time of quiet and contemplation. With fewer digital creatures to discover or chase than there were in the summer, the player is likely to spend their day admiring the leaves and making the rounds between all-too familiar landmarks from the previous seasons. If the player happens to approach the totem circle after dark, however, they will trigger an event that they had not yet encountered: the sky turns red, the stars begin to pulse, and puffy white orbs of mist begin to gather in the center of the totem circle as ominous music grows louder and louder. Suddenly, in a somewhat anticlimactic poof, a humanoid figure appears and immediately begins shuffling away, white mist trailing behind him. After moving away for a few yards, the figure turns, looks over its shoulder, and waves an arm, beckoning the player to follow as a playful melody accompanies its movements. This figure eventually leads players to the largest tree on the island, which lights up with an ethereal glow as it leaps up and disappears in its branches. In its place, the head of a spectral fox peeks out from behind another tree in the grove, prompting a game of hide-and-seek that lasts throughout the night.

For first-time players, this moment is meant to be a magical one, piquing their curiosity and encouraging them to seek out other secrets by revisiting old landmarks across the island. For someone who has already played *Proteus*, on the other hand, the existence of this autumnal encounter risks undermining the illusion of a living world that the game is attempting to convey. After learning the secret to summoning this spirit figure, the players gain knowledge of something they can exploit in this otherwise unexploitable environment: a scripted event they can trigger every autumn on any version of the island generated by the game. When combined with the too-clean colors, lively soundscapes, and fantastic creatures that could not exist in the actual world, *Proteus's* autumnal encounter can be framed as a well-crafted, repeatable, but ultimately hollow imitation of the wonder that can only be found in nature. What is more, this encounter reveals that, in order for this effect to trigger, the landmarks that seemed so idiosyncratic and unique to the player's particular experience—the totems, the tree, the cabin, the ruins, etc.—are actually prefabricated assets randomly assigned to different positions on every iteration of this island. As such, *Proteus* seems to reject a chance to focus on meaningfully local ecosystems in favor of generically spectacular ones, producing what Chang calls the "somewhat anaestheticizing sense that *this could be anywhere or nowhere at all*" (22, italics in original). Read this way, the beguilingly immersive "heterocosm" of *Proteus* becomes more like a comfortable "eco-friendly" hotel room than a living landscape: an inhabitable nonplace constructed from generic and carefully rearranged décor and whose insulated ecosystem threatens to distract its inhabitants from the world outside.

This rather cynical reading of *Proteus* highlights one of the issues with conceiving of games as "heterocosms." The term heterocosm—"literally an 'other world' or cosmos" as glossed by Hutcheon (14)—seems to imply an artificial alternative to the (natural) world usually inhabited by the player, one whose "'intensity of engagement' [is] unrivaled in most other media" and, by extension, in most other encounters with the natural world (51). Such a structural metaphor implies that we can only inhabit one world at a time; it is a small step away from the common refrain that the only function of games (or any piece of popular culture, for that matter) is as a temporary escape from the "actual" world, which puts them at odds with environmental activism by default. It is partially for this reason that Chang and other scholars of eco-games frame their objects of study not as heterocosms but as mesocosms: "experimental enclosures in size and complexity between small, highly controlled lab experiments and large, often unpredictable real-world environments," like sections of a field or body of water cordoned off for study (17). This model emphasizes not only the porous boundaries between digital games and the contexts in which they are played, but also draws attention to "the subtle negotiations that take place between human and nonhuman actors and technological assemblages during play" (18). In this sense, the concept of "mesocosm" actually resonates with Hutcheon's interpretatively productive concept of adaptations as "inherently palimpsestuous works" more strongly than the relatively closed notion of "heterocosm" does (6); mesocosms, like adaptations, are inherently connected to and affected by the natural/intertextual environments around them, operating not as mere copies of these environments but as spaces for experimentation, exploration, and interpretation of the environments they seek to re-create.

While the immersive spatial design of *Proteus* itself may encourage a reading of it as heterocosm, the intriguing paratexts of *Proteus: Artifact Edition* frame it as a particularly porous mesocosm with deep connections to the material world. This interpretation begins with the image on the front of the box (Fig. 17.1): beneath the game's title and framed in the jagged circular shape of the in-game moon are a set of four standing stones.

But these are not the mono-color, computer-generated stones one would see in the game; their smoothness, shading, and position in a bed of grass all point to this being a photograph of an actual place, a real-world referent lying just beneath the artificial reality of the game-world. The materials inside the box tell the same story: the back of the introductory booklet contains a photograph labeled "Frith Hall Ruins, Duddon Valley," depicting a structure that looks similar to the ruins that

Fig. 17.1 Cover art of *Proteus: Artifact Edition.* Photograph by author

sometimes appear on the island; the slipcase for the included audio disk features a photograph of a cabin nestled between hills beside a body of water, mimicking not only the game's lonely cabin but also the island's rolling topography; and a booklet titled "Field Guide and Walking Scores" even contains an unlabeled map of the Avebury Standing Stones in Wiltshire, England, adorned with a sketch of a spiraling portal in the center which evokes the mystical method of transitioning between seasons (Key 4). These images reveal that *Proteus* is not some dream of a nonexistent environment but is instead anchored in actual places that are meaningful to its designer, Ed Key. Just as games like *Marvel's Spider-Man* or *Assassin's Creed* condensed their real-world sources into navigable spaces through the process of selective fidelity, so too did Key bring together particular landmarks in order to adapt the experience of encountering them. Rather than faithfully re-creating any of the actual areas that surround these real-world sites—a task that would end up merely producing a mimetic heterocosm that would lose its sense of wonder immediately—Key's decision to procedurally generate the spaces between these evocative landmarks turns *Proteus* into a mesocosmic space of experimentation, playing with variables of distance, directionality, flora, and fauna to alter the relationships between these real-world spaces and the experience of discovery as one wanders between them.

The materials within the *Artifact Edition* do more than confirm real-world referents that affect the designer's experience, however; it also encourages the players to bridge the virtual and the material as well. Nowhere is this clearer than in the image on the front of the introductory booklet which details the game's development history (Fig. 17.2): a wisp of pixelated cloud stretches across the purple sky, slightly obscuring *Proteus* 's familiar moon. Around them are orbs of different sizes, looking a great deal like the pulsating stars that would occur during the autumn ritual described earlier in this chapter. The landscape silhouetted against this digitally rendered sky, however, looks to be real, even if it does resemble some of the topography of a Protean island: a gothic-looking tower and footbridge sits in the center of the image, flanked by a gently sloping hill to the left and a jagged line of trees to the right, each of which are reflected in the water below alongside the celestial bodies from the game. On the water's jagged edge (perhaps altered to mirror the generation of *Proteus's* shoreline) stands a figure who will be instantly recognizable to players as the one who beckons them onward during the aforementioned ritual, looking out at the scene with its back to the viewer. The long-winded label for this image provided in the back of this booklet—"Upper Neuadd Reservoir (March 2012) / Sky Over Home / Proteus night (Nov 2012) / Spirit figure"—confirms the blurred lines between the real and the virtual, the personal and the artificial, the

Fig. 17.2 "Upper Neuadd Reservoir (March 2012)/Sky Over Home / Proteus night (Nov 2012)/Spirit figure" in *Proteus Artifact Edition.* Photograph by author

temporally situated and the abstractly universal across which this image (and, indeed, the *Artifact Edition* as a whole) attempts to tread. The use of the spirit figure in this image—who looks apt to launch into his beckoning "follow me" animation at any moment—encourages the viewer to walk that line as well, perhaps by comparing what their own "Sky Over Home" might look like when composited with these images.

Beyond presenting an invitation to wander between the virtual and the material, *Proteus*: *Artifact Edition* actually provides a method for mapping it. Within a box labeled "The Protean Map" is a deck of cards, each labeled with a short statement that evokes an image of the natural world ("gathering clouds," "a cracked shell"), a shape or structure ("the great tower," "a skeletal frame"), or an abstract event or sensation ("an air of decay," "a traveler arrives"). While "Field Guide & Walking Scores" contains multiple ways to use this deck of cards, the rules for "Walking" are most relevant here:

- Choose an area to explore, shuffle the cards and begin walking
- Draw the top card. Find, create, or establish this feature in the world. If the feature is impossible, we may conclude that the journey described by the cards diverges from the physical world....
- Continue walking for a short time or a long time
- Draw a new card and repeat the process until you have walked enough or a [stop symbol] is drawn. (Key 26)

With these rules, *Proteus: Artifact Edition* has come full circle. The original digital game was designed by representing specific elements from the real world—landmarks, topographies, and the experience of wandering between them—in a vibrant virtual space. These are not mimetic representations of the real world, but fantastic objects with hidden secrets meant to communicate the wonder of exploring and experiencing the natural world. The "Walking" rules for the "Protean Map," in turn, invite the user to map the ethos of playful discovery and contemplation back onto the world around them, re-enchanting nature with the same sense of wonder that inspired the project in the first place. To be clear, this is not some augmented reality overlay like *Pokémon Go!* (Niantic, 2016) that populates the real world with pixelated creatures from *Proteus*—such a program would maintain the separations between real/virtual and nature/artifice that the *Artifact Edition* is trying to dissolve. Instead, by emphasizing the fluidity of the boundary between material and immaterial realities, the "Walking" rules encourage the user to look at nature with the same wonder-filled perspective practiced within the virtual space of the game. In this sense, *Proteus: Artifact Edition* is doing exactly what any good adaptation or environmentally oriented work ought to do: it encourages its users to approach the textual and/or natural "sources" that it emulates through the lens of an expressive interpretation and let the relationship between adapted and adapting spaces create new experiences all together.

Winter: An Open Conclusion

For all the sense of wonder and playfulness it brings into the world, there is also a more serious environmental message to *Proteus* that best comes through in the transition to winter. Unlike the soundscapes of previous seasons, the ambience of winter is nearly silent. All the sound has been reduced to a chilling high-pitched tone, occasionally punctuated with some odd echoes across the icy landscape. Whereas the weather in other seasons changes, winter always brings a low ceiling of slow-moving clouds, their grey color slightly offset from the dull white and pale blues of the snow and ice around. There are no creatures left in winter, no leaves on the twisted trees, no waving plants beneath the rows of standing stones that look more like grave markers in such bleak surroundings. All the player is able to do is wander through the dead landscape between the now inert landmarks until nightfall triggers the game's transcendent ending (which I will not spoil here). Suffice it to say, the game does not provide the player the opportunity to revert back to an earlier season; due to the statistical complexity of procedural generation, when the virtual eyes of the player-character close on the island of *Proteus,* the player will never see that particular island again. As opposed to other ecoadaptations that preserve a landscape in print or on celluloid, the procedural generation of *Proteus* captures the concept of ephemerality itself.

Such a melancholy ending presents two opposing interpretations to the player. On the one hand, it is tempting to read *Proteus* as using the cycle of seasons as an allegory for the youthful springs, halcyon summers, contemplative autumns, and lonely winters of a human life. In such a reading, the game offers players the fantasy of continual rebirth, exploring the delightful subtleties of "repetition with variation" that characterize both these regenerating islands and adaptations as a whole (Hutcheon 4). But in an ecocritical context, I cannot help but see something apocalyptic in *Proteus's* winter: it presents the prospect of a once vibrant and intimately familiar landscape dying on account of human (in)action. After all, whether it is due to a hunger for novelty and progress or an apathy for the once-vibrant and now too-familiar landscape around them, it is the player's choice to walk through the portal and move the world into its next season. The only difference is, once the metaphorical winter comes to *our* world, we will not have the choice to begin the game again. Regardless of which interpretation one is left with, one thing is clear: especially in the context of the goal-driven medium of video games—and, in fact, a goal-driven culture of endless productivity under late capitalism—*Proteus* and its *Artifact Edition* invite us to linger for a moment before rushing off to that next transformation, to look upon the world around us with fresh eyes and find, create, or establish the features we wish to see there as we wander along.

References

Buell, Lawrence. *The Environmental Imagination.* Harvard UP, 1995.

Castronova, Edward. *Synthetic Worlds.* U of Chicago P, 2005.

Chang, Alenda Y. *Playing Nature*. U of Minnesota P, 2019.

de Beke, Laura op, et al., editors. *Ecogames: Playful Perspectives on the Climate Crisis*. Amsterdam UP, 2024. JSTOR, doi.org/https://doi.org/10.2307/jj.10819591.

Dune: Awakening. Funcom, 2025.

Dune: Spice Wars. Shiro Games, 2022.

Fizek, Sonia. *Playing at a Distance: Borderlands of Video Game Aesthetic.* MIT P, 2022.

Gratton, Kyle. "Marvel's Spider-Man 2 Map Size Compared to the First Two Games." *Screenrant,* 20 October 2023, screenrant.com/marvels-spiderman-2-map-size-comparison-brooklyn-queens/. Accessed 31 January 2025.

Hogwarts: Legacy. Portkey Games, 2022.

Hutcheon, Linda. *A Theory of Adaptation.* 2006. 2nd ed., Routledge, 2013.

Kagen, Melissa. *Wandering Games.* MIT P, 2022.

Key, Ed. "Field Guide and Walking Scores." *Proteus, Artifact Edition,* Twisted Tree Games, 2016.

Louv, Richard. *Last Child in the Woods: Saving our Children from Nature-Deficit Disorder*. Algonquin Books, 2008.

Marvel's Spider-Man. Developed by Insomniac Games, Sony Interactive Entertainment, 2018.

Massongill, Justin. "The Tech Behind Marvel's Spiderman." *PlayStation Blog,* 6 September 2018, blog.playstation.com/2018/09/06/insomniac-interview-the-tech-behind-marvels-spider-man/. Accessed 31 January 2025.

Muir, John. *Steep Trails,* edited by William Frederick Badè, 1918. Ebook, Project Gutenberg, 1995.

O'Connor, Alice. "Self-Interviewing Devs: Proteus and 'Walking Simulators.'" *RockPaperShotgun,* 28 June 2014, www.rockpapershotgun.com/walking-simulator-proteus-ed-key-ricky-haggett. Accessed 31 January, 2025.

Proteus. Designed by Ed Key and David Kanaga, Twisted Tree Games, 2013.

Proteus: Artifact Edition. Designed by Ed Key and David Kanaga, Twisted Tree Games, 2016.

Tales of the Shire. Wētā Workshop, 2025.

Walden, a game. Designed by Tracy Fullerton, USC Game Innovation Lab, 2017.

Yang, Robert. "Level with Me, Ed Key." *RockPaperShotgun,* 29 November, 2011, www.rockpapershotgun.com/level-with-me-ed-key. Accessed 31 January 2025.

Index

P. Demory (ed.), *Ecoadaptation*, Palgrave Studies in Adaptation and Visual Culture,
https://doi.org/10.1007/978-3-032-09847-4